CONSTRUCTION TECHNIQUES IN SOUTH AND SOUTHEAST ASIA

HANDBOOK OF ORIENTAL STUDIES
HANDBUCH DER ORIENTALISTIK

SECTION THREE

SOUTH-EAST ASIA

EDITED BY

B. ARPS · M.C. RICKLEFS · D.K. WYATT

VOLUME FIFTEEN

CONSTRUCTION TECHNIQUES
IN SOUTH AND SOUTHEAST ASIA

CONSTRUCTION TECHNIQUES IN SOUTH AND SOUTHEAST ASIA

A History

BY

JACQUES DUMARÇAY

TRANSLATED BY

BARBARA SILVERSTONE
RAPHAËLLE DEDOURGE

TUTA SUB AEGIDE PALLAS · 1683 ·

BRILL
LEIDEN · BOSTON
2005

This book is printed on acid-free paper

Library of Congress Cataloging-in-Publication Data

Dumarçay, Jacques.
Construction techniques in South and Southeast Asia : a history / by Jacques Dumarçay ; translated by Barbara Silverstone and Raphaëlle Dedourge.
p. cm. — (Handbook of oriental studies. Section three, South-East Asia ; v. 15 = Handbuch der Orientalistik)
Translated from French.
Includes bibliographical references and index.
Contents: Layout and dimensions—Wood and carpentry—Wood, supports, and stiles—Mud as a construction material—Stone and stone cutting—Binders and plasterworks—Overhauls and repairs.
ISBN 90-04-14126-X
1. Building—Asia. 2. Building—Asia, Southeastern. I. Title. II. Handbuch der Orientalistik. Dritte Abteilung, Südostasien ; 15. Bd.

TH99.D86 2004
690'.095—dc22

2004058140

ISSN 0169-9571
ISBN 90 04 14126 X

PRINTED IN THE NETHERLANDS

CONTENTS

ACKNOWLEDGEMENTS

This work would never have been accomplished without the much appreciated help of my colleagues at the EFEO and the ongoing support of my wife Jacqueline Dumarçay, who proofread the French version. The English version would not exist without the careful work of two translators, Miss Raphaëlle Dedourge and Miss Barbara Silverstone, who were behind many modifications that improved the understanding of the text and, of course, the glossary. I give them all my warmest thanks.

INTRODUCTION

The history of techniques is that of difficulties overcome.
Claude Roy, *Paul Klee*, 1975.

Like all activities undertaken by man, construction requires access to a certain number of techniques that have become increasingly complicated with time. This study focuses on the development of this evolution. We have divided our work according to the materials used, each of which calls on its own particular methods. Obviously, the experience acquired with one material has been applied to other materials, which has often led to major mistakes. For example, when the Khmers switched from wood to sandstone, they continued for many years to use carpentry fitting methods, in particular, for their window frames. When they realized how fragile these frameworks were in stone, they did not stop using them completely. Instead, they kept these structures for the façades, while using frameworks that were better suited to the particular nature of sandstone inside the masonry (fig. 41 and 42).

We will touch on some points that go beyond pure technique, such as geometry. Geometry is an essential element in the art of construction, not only for staking out building sites, but also for the preparation of the materials. Stonecutting could not exist without geometry. We will also look at the strength of different materials. This has long been treated empirically, but it is a point that master builders have always taken into account.

In his book *Evolution et Technique*[1], André Leroi-Gouran has shown that the determinism of technique is as important as that of zoology. Thus, in the construction field, the fact that the shearing point of a horizontal wood beam is one-fifth of its length has determined the shape of what is called bending beam work. As long as this technique was respected, the techniques that followed could vary in formal detail. However, the rigour of the stress could never be ignored. In order for the bending beam work to remain stable, the stress must be located

---

[1] André Leroi-Gouran, *Evolution et Technique, Vol. I L'homme et la matière, Vol. II Milieu et techniques*. Albin Michel, Paris 1943-1945.

between the point of support and the shearing point. This is similar to the quarrying of stone. If quarrying has been abandoned despite the construction of monuments as large as the Borobudur, it is not so much due to the scarce quantity of blocks of stone of the required volume. It is rather due to the introduction of "stone standard forms" used with the commissions, which points to a certain normalisation of the volumes. It is the transformation of the worksite that has modified the technique of stone supplies and thus their mode of extraction. As we will see, improvements in stonecutting have in part given the authority to master builders to create architectural structures that break with tradition.

Leroi-Gouran has revealed how difficult it is to draw up a history of techniques. Thus, for the example of the bending beam work, it is impossible to say when this technique was first used. We only have a very general idea. The same can be said for the geographic location of its creation. In this sense, the title of this study must be moderated. It is not truly a history as the beginnings of architectural techniques are unknown and the location of their origins is generally vague. However, with the use of documents, we can reconstitute a history of their diffusion. Thus, for the example of the bending beam work again, we can rather precisely date their appearance in the Hindu states of Indochina. Working from there, we can deduce that their origin is Indian.

Architecture is defined as the art of construction. Yet, this definition ignores one aspect of this art: the anticipation of the finished work. The birth of a building is the carrying through of a project, as small as that may be. This can be said of all works, be they artistic or scientific. But this holds particularly true for architecture. The project expresses the desires of the master builder in the limits of his economic possibilities and technical knowledge at the moment. The object of this study is the history of these techniques in southern Asia a history of difficulties overcome.

The cultural area of Asia can be divided into three zones, each dominated by a particular civilisation: that of Iran, that of India and that of China. These are not areas sealed off from external influence. However, in each of these areas, builders were able to develop an architecture which possessed an original aesthetic and technique, despite its variety. The collective intelligence of each of these human groups who constructed in southern Asia gave the buildings of its location a particular shape that corresponded to its symbolic significance. This aspect has remained the same, despite technical evolutions. It not only holds true for the simplest buildings, but also for community and religious constructions. In southern Asia, under the dominant influence of India, religious symbolism imposed shapes and rituals which deeply marked architecture. This also resulted in the creation of treatises to simplify the respect of architectural rules and rituals. These texts attach little importance to actual construction techniques. They certainly do not take into account the new possibilities that their evolution made possible. This justified reproach accusing the treatises of freezing the architectural forms does not apply to southern Asia alone. It was also particularly rampant in Europe in the beginning of the 20th century (until about 1950) when the Vignole treatise, which was used in part as a basic teaching tool for architectural studies, was showered with a number of accusations.

Most likely in an attempt to escape this reproach, Indian rituals sometimes include basic technical data. For example, the *Rauravagama*, a treatise on ritual and Shivaistic doctrines, includes a chapter on various types of enclosures (chapter 41, Characteristics of enclosures). This section gives a basic description of proportions, but does not delve into their actual implementation. Here is the beginning of the chapter in the translation by B. Dagens and M.L. Barazer-Billoret: "I will briefly lay out the characteristics of enclosures. The three possible widths for the courtyard of the first enclosure are equal to, or two-thirds or one-half of the width of the temple. The measurements are taken from the base of the outer side of the pillars or the plinth, or even the orthostat. The diameter of the second enclosure is twice as large as that of the first enclosure..."[2]. Obviously, the perimeter of the courtyard cannot be created with such basic instructions. It is evident that the suggested dimensions are only meant as an indication, not as practical data. There is probably another reason for this. This treatise is mainly a ritual, and architecture is only an accessory of this. The construction site requires geometry: first for religious reasons, as geometry is a theophany[3], and second only for technical

[2] B. Dagens and M.L. Brazer-Billoret, *Le Rauravagama. Un traité de rituel et de doctrine Sivaites*, Institut français de Pondichéry, 2000; two volumes. Chapter 41 starts on page 249 of the first volume.

[3] Geometry as a proof of the existence of God is implicit, even though a form of atheism claims that geometry is older than the existence of the gods, who only submitted to the rules that existed before they themselves did.

reasons. To create a large courtyard such as one for a temple, it is necessary to use a device called an alidade. At first, this device may seem simple to use. However, it calls for exceptional knowledge. The precision of the results depends on the talent of the surveyor, which is difficult to foresee.

Sometimes the data indicated in the treatises are complex and seem to provide precisions that are only illusory. Thus the *Mayamata*, translated by B. Dagens, suggests the following procedure for tracing an octagon from a square: "To trace an octagon, the width of the (square) area being divided into twenty-four parts, a circle is drawn which extends over one part (beyond the sides of the area); then mediating lines are drawn from opposite sides (and bissectrices from siding corners); the octagon results from the joining, by the eight, (of the eight points thus determined"[4] fig. 1/a). This construction is given for an octagon of which four sides are part of the initial square. These constructions are designed to be undertaken on the exterior. The shape is constructed on the outside. In the sequel to their work, it seems that the authors abandoned this complication. For example, in reference to the shaping of the Linga[5], whose relationship with the initial square is similar to that cited below: "There are three ways (to draw an octagon) in a square whose width is that prescribed (for the Linga): the sought-after octagon may be drawn by taking the bissectrices from opposite angles (and the mediatrixes from opposite sides; (an octagon may, as well, be drawn) with a side equal to the mean between a quarter and a third (of the width of the initial square)". We can verify that three-sevenths of the side of the square are a good approximation of the side of the octagon, which entails a geometric construction that is slightly more complicated due to the division into seven equal parts of the side of the square (fig. 1/b). Also, despite a mediocre approximation, it is likely that the second proposition was chosen most often, the average between a quarter and a third of the side (fig. 1/C), except perhaps for the most prestigious buildings.

---

[4] *Mayamata. Traité sanscrit d'architecture. Edition critique.* Translation and notes by Bruno Dagens, Institut français d'Indologie. Pondichéry Vol. I, 1970, Vol. II, 1976.—Vol. I, p. 630 § 52 and fig. 30. Moreover, we know that the side of the octagon can be made using a circle drawn in a square of which the radius is equal to that of the circle circumscribed by the octagon. (fig. 1/d). This statement suffices to demonstrate that the addition of 1/24 on the side of the square does not make the construction any more precise.

[5] *Mayamata* op. cit. Vol. II p. 290 § 68-69.

This example of how to draw an octagon shows us to what extent geometry associated with architecture was poorly understood and how often builders settled for approximations. Thus, the construction of the streets of Mohenjo-Daro seems rigorous, yet it was done without a horizontal plane of projection[6], necessary when constructing on inclined land. This led to transforming the rectangle probably intended for the great bath into an irregular quadrilateral (fig. 2). These difficulties did not stop the vast undertakings that called for complex geometric work, such as the star surface of the Somenathpur temple, Karnataka, or the correction of axes as at Bapuon, Angkor, or the construction in an urban area of the enclosure of Angkor Thom.

Outside of the treatises, the techniques are rarely mentioned and more commonly they are referred to in the form of a metaphor. For example, the potter's work takes on a religious significance in the Khmer epigraph as on the stele of Pram Kuha Lûon, stanza I: "Victorious is Iça, who is moved neither by the desire to win nor the desire to lose, but whose desire, like a potter's wheel, shapes unceasingly all that moves"[7]. More than the potter's wheel, (represented on the foundation of the Shiva temple of Halebid at Karnataka, ph. 1), the movement caused by the beating of cream to create butter inspired a great many metaphors that referred to the creation of the world. For example, here is the text of stanza XCI of the inscription of Prasat Khna (Mlu Prei) which uses the beating metaphor for another conclusion: "Beat by the multitude of wise men with the mountain of a great problem, the milk ocean of its intelligence gave the desired meaning that had not yet been found"[8]. There are many examples of this myth. Some are works of art that provide no technical directions. However, in Borobudur, a carpenter is represented working with an adze on a relief on the first gallery. This does not seem to represent a metaphor, but rather construction work. Yet, when the bodhisattva carries the same tool on his shoulder on a relief on the fourth gallery (fig. 3), this is probably an allusion to the construction of the doctrinal structure of Buddhism. When they wanted to demonstrate on the reliefs of the exterior gallery of Bayon the amplitude of the work undertaken

[6] The creation of a horizontal surface before construction was not practised until 5 BC in Greece.

[7] G. Coedès, *Inscriptions du Cambodge*, Hanoi 1942, Vol. II, p. 12.

[8] G. Coedès, *Inscriptions du Cambodge*, Hanoi 1937, Vol. I, Inscription of Prasat Khna (Mlu Prei), stanza XCI, p. 216.

by King Jayavarman VII, the sculptors represented construction sites several times. On one of these panels (south gallery, west wing, fig. 4), an alidade is shown. In this case, this is not a metaphor, but a true work tool reproduced for itself.

In these few examples, we see to what extent the history of architectural techniques can only be studied through the actual structures and buildings themselves. The data of the texts and graphic documents, as interesting as they may be, are too fragmentary. The information in the treatises would probably have stunted progress had they been more explicit and especially, more closely associated with a certain type of architecture. There was a normal gap between the appearance of the technique, its generalised use and especially the attempts of normalising the treatises. This is particularly noticeable with stonecutting and carpentry. Thus, the treatise of the *Mayamata*[9] gives imprecise technical instructions for the construction of the radiating beam work on a square structure. The treatise also features omissions that seem to take for granted that all the rafters come from the core, which of course is impossible. The rigour of the calculation of the length of these rafters is also illusory (we will look at this text in more detail in chapter II). In this case, the treatise only serves as a sort of basic guideline for the master builder. It indicates that the rafters of a framework encircling a square structure are unequal in length and gives the master builder a graphic calculation method that can only be rigorously applied for very few. In Bali[10], an effort was made to update the treatises (for example by introducing a few ideas about concrete). But in practice, these directions are not respected. The texts are accessible to everyone, but only few people are able to read them. So, when a master builder wants to know if his project is in accordance with the rituals, he asks the person "who knows". In other words, despite the few efforts that have been taken to update the texts, tradition has the last word—or more exactly, whatever remains of it. For example, in the Balinese treatises, the Hasta Kosali, which are probably of Javanese inspiration, we can see instructions on the height of doors. We can also remark

---

[9] B. Dagens, *Mayamata*. Vol. I, p. 368, note 23 and fig. 19.

[10] In his article "An introduction to the cultural study of traditional Balinese architecture", L.E. Howe, Archipel No. 25, 1983, makes a similar remark without mentioning the gap between the attempts of modernising the texts. He says: The texts are then only consulted when there is some disagreement about rules or procedure.

that on several Javanese temples of the 13th century, these instructions were respected. But this is not the case in Bali (we will talk more about these dimensions in the third chapter).

The representation of an alidade on a relief of Bayon—as interesting as it is in that it is a concrete proof of the tool's use, which otherwise would be unknown to us except for the traces left on monuments—is in no way an explanation of its use. Only the marks left by the (land) surveyors on the monuments have allowed us to reconstitute the construction methods which, all the same, are the subject of many chapters in the treatises.

The study of the techniques, based on the actual constructed works, is difficult because in principle the work as we see it is finished and thus the technical process should not be discernable. Yet it is partly visible, because the monuments often remain largely unfinished. For example, the planned restoration of stone monuments—often a huge task—was started but never finished. We see by this example that there is another problem; the measurements are not exact because they are only definitive after the restoration, which often reaches very large proportions. On the Bayon of Angkor, for example, these proportions reach up to 0.25 m in width.

The first contacts with the Western world did not bring about notable transformations in technique. The Greeks of Alexander and the kingdom of Bactria profoundly influenced sculpture and architectural décor, but had little effect on construction techniques. When we began to better understand European architecture, particularly thanks to architectural treatises, this did not cause major change. Thus, much later in the 18th century, when the treatises of Palladio and Serlio were introduced to the Javanese, this did not bring about any change in their form of roofing; the climatic conditions did not allow for it. When the European occupation took place, the benefits of the triangulated framework were put into place and its use became more generalised. However, as we have observed for the straight ridge poles which were transformed without changing the exterior appearance, the buildings with roofing set on radiating beam work in Java and Bali now had a triangulated framework with rafters perpendicular to the wall plates and an unchanged external appearance.

Thus it seems that, paradoxically enough, the history of construction techniques in southern Asia only exists independently from the evolution of architecture.

CHAPTER ONE

LAYOUT AND DIMENSIONS

Once the project design is finished, the preparation of the work completed and the master builder has accepted the dimensions adapted to the use of the building, the first technical operation is that of laying out the terrain. Geometry plays an important role in this first moment of constructing a building, not only because it is a proof of the existence of divine powers but also because it integrates the future structure into its environment, which itself contains geometric shapes used by the Creator.

As early as period IV of Mundigak and the civilisation of the Indus, builders oriented structures in their natural environment. For these ancient eras, it is difficult to reconstitute the techniques; we can only observe the results. The streets of Mohenjo-Daro are positioned, as is the silo of Harappa. Nevertheless, when a building was sited on an inclined terrain, e.g. at Mohenjo-Daro for a building as elaborate as the great bathhouse, the builders did not yet know how to create a horizontal plane of projection. Thus, the bathhouse seems to be built on an irregular quadrilateral (the E, F, G, H rectangle of fig. 2 following the layout of J. Marshall[11]). The marking out was thus probably done directly on the natural slope of the terrain as this would facilitate water evacuation (K of fig. 2), and as the layout was probably conducted with a set square (the A, B, C, D rectangle of fig. 2 reconstitutes the staking-out of the ground). This allows us to reconstitute the slope of the terrain at the time of layout (the L, M line of fig. 2). As the building was gradually constructed, the builders either did not know how to correct the problem or did not want to.

The architectural treatises of India pointed out the definition of the cardinal points of a space and their placement. The following is an example of such a text taken from the *Mayamata* from the translation by B. Dagens: "when the gnomon has been made it is set up in the chosen place at sunrise, then a circle is drawn of which the gnomon

[11] J. Marshall, *Mohenjo-Daro and the Indus Civilization*, London 1931.

is the centre and of which the diameter is double the length of the gnomon.

"The line which joins the two points where the shadow (of the gnomon) has touched the circle, in the morning and in the evening, gives the east-west direction. The line which passes through the space between these two points and (which is like that which) connects the head and tail of a carp, is the north-south axis"[12].

In another treatise, the *Manasara*[13], contemporary with the Cola dynasty, the method is similar. In the Balinese treatises *Hasta Kosali*, identical instructions are mentioned. Nevertheless, in the contemporary practice, they go about the process differently. They put a stake vertically into the ground and trace a circle around it (of any diameter, but always less than the height of the stake). The shadow of the stake crosses the circle in the morning and in the evening. The straight line joining these two points is the north-south axis of the terrain and its perpendicular is the east-west axis.

By erecting the perpendiculars on these axes, the builders were able to determine the corners of the building. We can interpret the words of the *Mayamata* cited above, "the head and tail of the carp", as an allusion to the two arcs necessary to erect a perpendicular. It is also on these axes that points are fixed, which make it possible as the building is gradually erected to preserve the axes by their protraction using the alidade. Although today, the limits that must have materialised these points were never discovered, traces of protraction are highly present in Java and Cambodia.

The layout is often dependent on factors in domains other than architecture. Thus, Borobudur of the 8th and 9th centuries is today seen as a coherent Buddhist monument. Yet it has a very complex construction due to its numerous architectural renovations. At first (W of fig. 5), the monument had four axes of symmetry to fulfil its first purpose, which was Hindu. After the renovations by the Buddhists, the new master builder imposed a new layout that featured only one axis of symmetry (X of fig. 5). At the top, a basic structure set on a terrace features a deformed circle structure (Y of fig. 5). After the collapse of the upper section, this last design was taken up on three layered terraces (Z of fig. 5). This last renovation was accompanied by a major stabilisation that features a layout that follows that of the

[12] B. Dagens, *Mayamata* op. cit., pp. 11-12.
[13] P.K. Acharya, *Manasara* T. IV, pp. 24-25.

original monument (V of fig. 5). This last work was itself modified, but with the same layout (U of fig. 5).

The excavations of Candi Sewu[14] overseen by the archaeological department of Indonesia made it possible to determine the numerous steps of the staking-out of the central building of this immense sanctuary. After the axes were determined, a square measuring 41 meters on each side was drawn on the approximately horizontal ground. The surface of this square was dug up in an irregular fashion and freed of any rocks that filled it. The excavation was then banked up with alternate layers of sand and gravel up to 25 cm from the surface. On the upper layer of the backfill, a large mandala was traced, then covered by 60 cm of alternate layers of backfill. On the upper surface of this backfill, at the centre, a new square measuring 5.29 meters on each side was traced out. Its corners were marked by parallelepipedic boundary stones. On this square, the builders erected the brick altar with its nineteen courses following the Vedic rites (the dotted line of fig. 6 indicates the upper level of the altar). They redefined the square of the base of which the diagonals and axes were traced (fig. 6) on the level of the ninth course. It is on this base that the very large pedestal and central statue were erected before any other construction took place. Only after this procedure did the construction of the temple begin.

In India in the 10th and 11th centuries, construction layouts met with these same criteria. For example, the great temples of Tanjavur and Gangaikondacholapuram are, on the one hand, copies of another building. The axes of the actually constructed temple and those of the featured construction are the same. However, in the siting of these great temples, the builders no doubt took into account the vision that they must have had of the finished structure so that the image would be coherent. This implies a cleaning down of the viewpoints where the designed building would seem to be expressed correctly. Thus the layout corresponds to a complex plan that is not only geometrical but that also takes into consideration the perspective axis.

From the 12th century onward, the master builders disassociated the layout of the real building from that of the designed building. The treatises that we have consulted do not point to the presence of a different ritual according to whether the architecture was a real volume

[14] J. Dumarçay, "Le démontage du temple central du Candi Sewu" *B.E.F.E.O.* *O.* LXXVI. Paris 1987, pp. 289-310.

or an element of a designed building of which we only see a part. For a single construction, the builders stake out not only the building that they will actually construct, but also the designed one. For example, the north-south axis of the principal temple of Darasuram which is truly constructed and that of the figurative element are parallel but different because a part of the projecting part of the temple is covered by the principal tower shared by both buildings (fig. 7). We can observe the same layout principles at Thribuvanam (constructed between 1178 and 1218, not far from Darasuram). This shows, were it necessary to do so, the extent to which the initial project was laid-out and to what extent the image that the builders wanted to give to the structure was an integral part of the constructed architectural work. It is thus probable that the technique was not repeated and that, for the master builder, the structure maintained its unity.

The sanctuary towers of the great temples of India are comprised of numerous false floors of which the structure sometimes differs remarkably from the structure of the body of the main building. It was thus necessary to create new sitings. This is the case of the temple of Gangaikondacholapuram[15]. The base of this temple is square. This is also the case of the first false floor but the second floor features a structure that is partly octagonal, which hints at complex geometric work. It is unlikely that this work was done on the monument itself, but rather on the ground and implemented once the stonecutting was finished. Four sides of the octagon are parallel to the initial square. But before the principal structure of the square structure, the implementation was thus developed from the square of the foundation of the false floor (square ABCD of our fig. 8, the third false floor). It is likely that the most elaborate process in the *Mayamata* (which we described in the introduction, see fig. 1/b) was used and that the protraction of the principal axes alone allowed for the implementation of the structure once the groundwork was laid. We can verify the dimensions of the fourth, fifth and sixth false floors. We can also note that the values obtained using the 3/7th method as shown in the *Mayamata* are all close to reality.

The protraction of the axes, effected as the building was gradually

[15] For all information concerning the temple of Gangaikondacholapuram, we consulted the work of Pierre Pichard, *Vingt ans après Tanjavur, Gangaikondacholapuram*, Mémoires archéologiques 20 de l'EFEO, Paris 1994. For the layout of the false floors of this temple, Volume I, p. 89-91 and Volume II, P.1 17-19.

erected, can be observed on a number of monuments. In Cambodia, when the temple of Bapuon was torn down to be renovated, we were able to observe the protraction of the axes engraved on the stones inserted in the internal backfill. In Java, in the central court of the temple of Prambanam, nine boundary stones were fit into the backfill and were engraved with the trace of the axes (fig. 9). As the construction gradually progressed, these boundary stones were covered with new boundary stones that bore the same indications. At the end, the last boundary stone was topped with a small linga; the ensemble was set in a small shelter in the form of a little temple (on fig. 9, the little temples indicated with the letter x are solid; for the other structures, the arrows indicate the direction of the opening) which protected the tracing of the axes and the diagonals of the enclosure.

Even so, the layout is not complete. It is taken up before the cleaning can begin. It is only during this last procedure that the definitive dimensions are given, those corresponding to the project and to its significance. On the temple of Bayon on which the cleaning was not finished, in particular on the internal walls of the galleries that encircle the inner courts, we can observe that the work was planned. For this, the axes of these structures were protracted. They were materialized by the letter "kaf" of the ancient alphabet, which probably signifies the word "kandal", the middle, the centre[16]—the letter most often engraved on the inner side of the real lintel of the doors. We see this letter on many Khmer monuments of the 12th century, in particular to indicate the axis of stairways. It was also used by sculptors (Ph. 2), generally to show more or less clearly the axis of the panels of the bas-reliefs.

Dimensions

When he arrived in India in 1760, Father Coeurdoux of the Missions Etrangères noticed that it was completely natural for Indians to apply the division of time to that of space.[17] According to Father Coeurdoux, this usage was widespread, from Cape Comorin to the

---

[16] Interpretation of B.P. Groslier, *Le Bayon* op. cit. p. 25 note 3.

[17] *Lettres édifiantes et curieuses écrites des Missions étrangères.* Mémoires des Indes, Lyon éditions de 1819. Volume VIII. P. 338. Letter from Father Coeurdoux to Mr. Delisle of the Academy of Sciences on the itinerary measures used in Eastern India. Pondichéry, 12 February 1760.

far ends of India, by all the populations of the country. This shows to what extent the measurements are imprecise. A league was considered to be the space covered by a man walking neither too fast nor too slow during a vigil (about three hours). However, in India and in the Hindu countries, there were major works that involve fairly precise measurements. For example, the ramparts of Angkor Thom, despite a very dense occupation of the space, form a correct square measuring on the east side 3029.6 metres, the north side 3089 metres, the west 3037.8 metres and the south 3050.3 metres[18], which averages out to be 3051.7 metres per side. The biggest variation is on the north with about 40 metres, due seemingly to the error of an angular layout which shows one grade south-west. When the construction project of the rampart was established, these dimensions were probably calculated from a unit created for this structure. It is likely that at Angkor in the beginning of the 13th century when the wall of Angkor Thom was erected, there was no general measurement system.

The same can be said of monumental architecture. An example would be the Borobudur whose dimensions seem rigorous. However, when we attempted to establish the reconstruction plans, we came across several difficulties at the time of restoration. Due to the internal thrust of the backfill, the walls of the first gallery were slanted toward the exterior, which noticeably widened the dimensions. So, we chose a reference line that we assumed to be horizontal and rotated it to obtain the dimensions of the monument, as the walls were vertical. The variations between the collected measurements of the ruined monument and the values reconstituted by the rotation are often quite remarkable. For example, panel D of the north side of the first gallery has a length, in a slanted position, of 2.72 meters, which after rotation comes to only 2.48 meters. The same goes for the east side: panel A measures 12.47 m in length, and only 12.33 meters after rotation[19]. The architectural history of the monument, on account of its change from Hinduism to Buddhism, attests to major modifications. In particular, the perspective effects disappeared when the width of the stairs was made consistent. We realise that under these conditions, the foot or cubit used in the construction cannot be reconstituted with

---

[18] A.Y. Bosco, *Situation topographique du Bayon*, annex IV by J. Dumarçay, *Le Bayon, Histoire architecturale du temple*. P. 74 and P. 1 XXI.

[19] The entire set of these measurements was published in J. Dumarçay, Report on measurements and dimensions, *Pelita Borobudur Seri CC No. 3*, Jakarta 1982; pp. 225-234.

precision. Many consider the measurements of the monument to be dimensions of the circumstances that adapted a former construction to a new project. In addition to these uncertainties, there are the techniques to consider. Between the layout and the application of the plaster, there was another important construction step: restoration. This was sometimes a large-scale operation. It reached up to 15 cm on a stone building in Angkor Wat and even more in Bayon. On a brick monument, the thickness of the plaster prevents us from taking a rigorous value of the building's dimensions. At Preah Koh of Roluos, the sanctuary towers kept a part of their plaster, which was applied in several successive layers. The last layer is composed of a thick plaster which was modelled (on Ph. 3 on the right pier, several layers of superimposed plaster can be seen). This reaches 6 centimetres, and even more in some areas.

Concerning the civilisation of the Indus (with the base of 150 measurements), we tried to determine the units that were used in staking out these structures, in particular at Harappa and Mohenjo-Daro. We discovered two values that might have been used together: cubits (whose value can vary from 51.56 cm to 52.83 cm) and feet (which vary from 33.02 to 33.52 cm)[20]. We converted the dimensions of the great bath of Mohenjo-Daro. With these measures, we came to a result that seems satisfying: 34.9x20.58 for a foot of 33.52 cm and 35.4x20.89 for a foot of 33.02 cm. But this does not take into account the fact that the monument was sited on inclined terrain and that the land surveyor did not project the measures on the inclined land. We can note that the dimensions are noticeably more satisfying, as the reconstituted layout is 11.4 x 8.4 meters, which represents 34x25 for a foot of 33.52 cm or 34.5x25.4 for a foot of 33.02 cm. These latter indications reveal the extent to which it is difficult to calculate the desired dimensions for a construction as elaborate as a great bath. For many other structures, we can only approximate.

The dimensions of the materials planned for construction are set according to their use. For example, if we intend to build a wall with alternating header and tile bricks, their length must be double their width.

Dimensions are difficult to establish correctly. Something that seems the most obvious and the easiest, e.g. the dimension of bricks, as they

[20] M. Wheeler, *The Indus civilization*, University Press, Cambridge, 1960, p. 66. J.M. Casal, *La civilisation de l'Indus et ses énigmes*. Paris, Fayard, 1969, pp. 126-127.

are created from a single mould for the same structure, actually produces a great variety of sizes due to reasons of drying or carelessness. Here for example, the bricks of a generally square shape of the Kushan era collected by Marc Le Berre in Bactria[21] near Arg, on the exterior side of the rampart (period 1) with dimensions expressed in meters:

0.310 x 0.300 x 0.125
0.310 x 0.300 x 0.140
0.330 x 0.310 x 0.110
0.340 x 0.320 x 0.130
0.330 x 0.310 x 0.135
0.325 x 0.320 x 0.140

Also taken by M. Le Berre, but on the eastern rampart of Bactria (period 1 A):

0.450 x 0.440 x 0.140
0.440 x 0.435 x 0.140
0.450 x 0.440 x 0.130
0.450 x 0.440 x 0.150
0.450 x 0.435 x 0.150
0.445 x 0.440 x 0.150

M. Le Berre also collected bricks from the stupa on Top-I Roustam. These are baked bricks of a trapezoidal form. The dimensions are given as follows: length, big side, small side, thickness.

0.550 x 0.300 x 0.295 x 0.080
0.550 x 0.305 x 0.300 x 0.080
0.550 x 0.300 x 0.295 x 0.080
0.560 x 0.300 x 0.275 x 0.080
0.570 x 0.300 x 0.275 x 0.080
0.555 x 0.300 x 0.300 x 0.080

The dimensions of the large square unburnt bricks, regardless of their placement, are irregular. On the other hand, the stupa bricks, despite the complication of their shape, which must take into account the hemispheric structure of the building, are regular. At Bactria, we

[21] Bruno Dagens, Marc Le Berre and Daniel Sclumberger, *Monuments Préislamiques d'Afghanistan*, Mémoires de la Délégation archéologique française en Afghanistan, Vol. XIX; third section "Observations sur les remparts de Bactria" by M. Le Berre and D. Sclumberger, pp. 92 to 102. The dimensions measured being quite abundant, we will only reproduce a few here.

can observe what is a common rule: the more difficulties a worksite is faced with, the better its layout and the protraction of the dimensions are, even for the bricks.

The protraction of measurements has always been a difficult problem to solve. The Candi Sewu temple was built in Java at the end of the 8th century. It comprises a great number of similar chapels (240), so in principle the dimensions should be the same. Starting from this observation, Pascal Lordereau undertook a study that made it possible to determine the cubit[22], basing his work on the great number of buildings that were theoretically similar. Lordereau proceeded in the following manner. First, limiting his study to the chapels of the first level of which he measured certain elements, measuring a maximum measurement, such as we see today with all the joints open, and a minimum measurement with all the joints closed. Remember that the structures of the Candi Sewu are built with joints open, which makes it possible to reconstitute the measurements with the joints closed. Thanks to considerable statistical work, Lordereau was able to determine that a cubit measured 0.348 meters, which leads to whole values for the principal measurements.

Then, Lordereau attempted to apply the cubit to Borobudur[23]. The results, as interesting as they are, are difficult to use due to the architectural history of the monument.

The values measured on a Khmer monument as elaborate as Pre-Rup show that the protraction was carefully executed (despite some errors), indicated in the following table which concerns the internal angles of the stairs that lead to the first terrace (the letters of the table refer to fig. 10). By calculating the value of the hypotenuse as if the layout was completely orthogonal, we see that the errors are minimal. The differences are indicated between parentheses in the following table. If the angles truly measured 90°, (the values of the following table are shown in meters):

A-B 5.14; B-C 18.97; A-C 19.86 (-0.21)
C-D 20.30; D-E 6.06; E-C 21.18 (=)
F-G 6.08; H-G 18.95; F-H 19.69 (+0.21)

---

22 Pascal Lordereau, La coudée Indo-Javanaise; in J. Dumarçay *Candi Sewu et l'architecture bouddhique du centre de Java*, EFEO, Paris 1981; pp. 45-73. The indications that follow concerning the chapels of Candi Sewu are mostly taken from this work.

23 A precedent study was undertaken by Jacques Ducamp, Etude numérique des formes du Borobudur in J. Dumarçay , Histoire architecturale du Borobudur, EFEO Paris 1977, Appendix II, pp. 73-77.

H-I 19.00; I-J 5.05; J-H 19.62 (+0.03)
K-L 5.10; L-M 19.60; M-K 19.90 (-0.23)
M-N 19.00; N-O 6.05; O-M 19.93 (-0.03)
Q-P 6.02; Q-R 20.32; P-R 21.14 (+0.05)
R-S 18.93; S-T 5.94; R-T 19.97 (-0.13)

We can observe that the symmetry is respected. Thus, the values of lengths B-C and H-I correspond almost perfectly. The same can be said of H-G corresponding to M-N. This is also the case of L-M and S-T, as well as Q-R and D-C, and this despite the particular treatment of the eastern façade of the monument and the modifications made to the outer steps. The layout seems to show a remarkable precision (the variations of the hypotenuse do not go over 23 cm). Yet, despite this rigour, it seems difficult to reconstitute the model that was used.

While the protraction of the dimensions is satisfying at Pre Rup, this is not the case at Sumatra for the Muara Takus monuments, which were measured with great care by Indonesian archaeologists[24].

The following table reproduces a part of table No. 15, indicating the dimensions of Candi Tua, from the Indonesian publication. The letters refer to those of our figure 11 (the values of the following table are indicated in meters).

	Foundation I	Foundation II	Foundation III
a	2.57	2.20	4.32
b	1.30	1.07	0.74
c	2.55	2.30	0.75
d	1.83	1.65	0.78
e	2.57	2.20	0.75
f	2.15	2.10	0.75
g	1.73	1.03	0.76
h	2.45	2.18	0.77
i	1.73	1.03	0.74
j	8.85	7.40	4.10
k	3.08	3.05	-

[24] The measurements of the table are taken from Ismiyono, Mulyono, Bambang Sumedi, Bambang Siswoyo and Winarto, *Laporan Studi Teknis Muara Takus, Bidang Tehno Arkeologi. 1983.* Direktorat Perlindurian dan Pembinaan Peningalan Sejarah dan Purbakala 1983. The diagram of our figure 10 reproduces in part figure 13 of the report as well as the attached table.

We can observe that the ratio of the dimensions between the two foundations I and II is not consistent. It varies from 1.10 to 1.67 meters. It is possible to multiply the examples. Even for very simple buildings, references were made to the owner's finger, which is a way of showing that the measurements are not based on a fixed unit before the establishment of the project, but to a module exclusive to the project. We might say that in southern Asia, there are as many measurement systems as there are monuments. As the treatises indicate, these are not metric relations, but proportional ones.

Pierre Pichard, in his study on Gangaikondacholapuram, perfectly highlighted the use of the module on this temple with its very elaborate structure. The variations are minimal between the theoretical dimensions and the measured values[25]. Thus, P. Pichard was able to establish that the monument had been constructed from a module of 38.64 cm, and after multiple verifications, he concluded that: "the difference between the theoretical dimensions calculated on the basis of 38.64 cm, and the dimensions truly measured on the southern façade of the sanctuary tower always remains less than 2 cm".

Chapter II of the *Manasara*[26] deals with measurements whose point of departure, after the atom and imperceptible measurements, is as follows: § 48: "Eight hair-ends joined together make what is called a *liksha* (nit); § 49: Eight nits combined together are called a *yuka* (louse)". We can see to what extent this point of departure makes it possible to give any value to the base of the measurement system proposed in the treatise. We find the same type of point of departure in the *Hasta Kosali*[27]: here, it is the first phalanx of the index finger, the "guli", the following measurements are multiples of this unit. As the values indicated in the treatise are indicated in "guli", in particular for the cornice outline, we realise that if the base is too narrow, the mouldings become imperceptible, which leads to simplification[28]. The cornice outline holds a very important place in the *Mayamata*, with a large number of moulding series that cannot be reconstituted unless simplified. This is due to the fact that the absolute value of the mod-

---

[25] P. Pichard, *Gangaikondacholapuram*, op. cit. Vol. I, pp. 74, 75 and 77.

[26] *Manasara* op. cit., pp. 5-9

[27] *Hasta Kosali* op. cit.

[28] Obviously, this does not apply to southern Asia alone. In Turkey for example, a fathom is often used, which must be equal to the length of two stretched-out arms. This leads to variations despite the official value of the fathom at 1.82 meters.

ule can vary widely from one monument to the next. These extracts from chapter XIV "The foundation" § 11b-16 are a good example; they concern the height of the foundation according to the class of the occupier. "For the gods that height is four cubits, for brahminsit three and a half cubits, for kings three cubits, for crown princes two and a half, for merchants two and lastly, for sudra, one cubit". But the foundation can be calculated according to the architecture depending on the number of storeys: "starting with the height of one pole (i.e. four cubits) for the base of a building with twelve storeys and decreasing by six digits (for each storey less) down to buildings with three storeys, the largest of which have a base one and three quarter cubits high. Another method of calculating the height of the base is indicated by the sages for smaller buildings: this height should be equal to half that of the corresponding pillar less a sixth or an eighth".

If we consider this text, the value of the fathom is equal to 54 fingers (6 x 9) + 1 cubit ¾. This is consistent with the value of the fathom indicated in chapter V where the author of the *Mayamata* indicates that a cubit is equal to 24 fingers, and four cubits equal a fathom, which we can indeed verify: 54+24+18=96 and 96/24=4. The consistency of these latter calculations should not be deceptive as it is based on completely random data: the value that we give to a finger. Moreover, the author is certainly aware of this rigour that he recommends, while at the same time suggesting the use of the fathom or cubit according to the volume to be measured.

We can see that for the master builders of southern Asia, whether they used treatises or monumental savoir-faire, there was no coherent measurement system, but rather a proportional one that existed for each structure: a modular system.

CHAPTER TWO

WOOD AND CARPENTRY

There is no doubt that wood is the most commonly used material in southern Asia. Almost every species has been used, even those that would seem the least likely to ensure good results. For example, date palm trunks were used for carpentry at Baluchistan after a rough scantling by splitting the wood with wedges. In Cambodia and Java, the trunks of coconut palms are trimmed using the same technique. This method is demonstrated on a relief of Bayon (outer gallery, southern side, west wing, Ph. 4). The poplar tree is probably the only wood used today for purlins in northern India and Pakistan. In regions with large forests, it is not only the quality of the wood, but also its market value—sometimes completely independent of the qualities required for construction—that justify its use. Thus, a trunk of teak wood (*Tectona grandis*) keeps its value and is used even if the wood is flawed. In Thailand, for example, we can see crooked pagoda pillars that are made of teak, which is reserved for the noblest buildings, pagodas and palaces. We are thus dealing with a quality independent of technical value. In Cambodia, the *Hopea odorata* ("koki" in Khmer) plays the same role. It is a very dense wood (about 0.80) that is highly resistant against insects and has low flammability properties. In the 1960s, in the Siem Reap region of Cambodia, when a pagoda was under construction, a plantation of "koki" was planted at the same time. The "koki" requires forty years to reach maturity, and as this species was beginning to disappear in the forest, they had to be sure to be able to replace flawed beams or pillars if necessary. For other buildings, the builders used other varieties of *Dipterocarpaceae*, which unfortunately proved highly flammable[29]. Many towns and villages burned down, and arsonists (even when the fires were set on accident) were severely punished. For this reason, the kitchen is separated from the house and is usually set on the ground. In this region, the *Shorea obtusas* was also used often. It is very dense (1.08) and highly resistant

[29] The resin of one variety of Dipterocarpaceae (*Dipterocarpaceae alatus*) is used to caulk small boats and make torches.

against insects. This wood is also used for inland fleets, especially in Java where the *Shorea acuminata* is used. Finally, in southern Asia, all varieties of bamboo are used, as well as all sorts of leguminous plants, which were sometimes given a value independent of their quality. Thus in Sulawesi, the *Drosperos celebrica* (a type of extremely rare ebony) is thought to ensure extra protection for the inhabitants of a house constructed partly of an expensive wood of mediocre quality.

Before their use on the worksites, the trunks often undergo a preparation called retting. This consists of soaking the wood in a muddy vat. The wood is prevented from floating on the surface of the water and a system of transverse beams keeps all of the wood underwater. The same process is used for the beams of the framework, when the roofing material is plant-based material. The process is more or less long; the biggest beams may undergo the treatment for several years. For roofing set over screeds, it was necessary to soak wood in water in order to curve the beams, which called for a long preparation before the construction could begin. This probably explains why large shafts were rare, and why the technique of roofing over screeds for large buildings fell into disuse in about the 9th century. However, the use of curved roofing for small constructions was used for a great deal longer, in particular for rice silos (in Bali and Lombok).

Woodworking was largely dependent on the evolution of tools. The axe and the adze have been used since the Stone Age. Each of these tools was found in Mundigak in Afghanistan, made of bronze and featuring a hole for a handle. These two tools, along with scissors, were the sole tools of carpenters for a very long time. A carpenter using an adze is represented on a relief of Borobudur. Contemporary workmen of Java, even today, often have only an adze as their work tool, which is used with great dexterity (the axe being used only by woodcutters). Although the saw has been used since the Bronze Age[30], it seems that it only appeared in India at the beginning of the Christian era, perhaps under the influence of the Romans. It is difficult to pinpoint the date that these tools arrived in south-eastern Asia. The representation of a scene depicting wood being trimmed by splitting in the 13th century (Ph. 4) is not a good indication of the rapid diffusion of this tool. However, taking into account the creation of complex stiles as early as the 7th century, it is likely that the saw

---

[30] These notes on the origin of woodworking tools were taken from W.L. Goodman, *A History of Woodworking Tools*. G. Bell and sons LTD. London, 1964.

existed but was little used at that time. Plugs are now widely used by carpenters in southern Asia. It is difficult to say when the drill was introduced in the region; the "arc drill" has existed in Europe since the Roman era[31]. In Java, Sumatra, Bali and Lombok, workers used a drill known as "Archimedes" as it was manufactured with the principle of an endless screw (although despite its name, this tool does not seem to have been commonly used before the middle of the 19th century). But it was created as early as this era in Indonesia, where artisans gave it an appealing shape (Ph. 5). For the tracings and especially the alignment of the trusses, the carpenters used blackened rope rolled around a winch (a very common practice). This was probably created around the 18th century, and is depicted on a fresco in the Wat Buak Klok Luang in Chiang Mai (fig. 12). The artisans also gave this tool a beautiful form (Ph. 5, bottom).

The assembly of the wooden structures often consisted of simple layering with no embedding. For example, the floors of different storeys of the pavilions at the entry to the temple of Chindambaran rest on joists placed over the beams. A small piece of wood tops the pillar simply to increase the support area. Nevertheless, the assemblies of beams are quite numerous. Those which we were able to reconstitute by using the imprints left in the stone were part-wood assemblies. As the large vats gradually became rarer, builders created assemblies that were sometimes very complex to ensure the rigidity of the entire structure. Chapter XVII of the *Mayamata* covers this subject. It seems that it recommends only part-wood assemblies, even for vertical pieces. In this text, tenons and grooves are barely mentioned. However, one paragraph makes note of the dovetail tenon. This form is used for stone construction; its transposition to wood architecture is difficult. Chapter XVII of the *Manasara* also mentions assemblies with indications that are somewhat troubling: "The death of the master would occur if the nail of joints was fixed to the middle of the pillar in the centre of the house"[32]. In stone buildings, the wall plate set in the cornice sometimes supported only the trusses. The base of the rafters is attached to the upper level of the cornice (Ph. 7).

It is most likely that it was after the end of the 18th century that assemblies more complex than part-wood ones came into use. These

---

[31] Goodman op. cit. p. 160.

[32] Prassanna Kumar Acharya, *Architecture of Manasara, translated from Original Sanskrit, Manasara Series Vol IV.* Ch. XVII. § 205-206. p. 198.

were reserved for horizontal beams (ridge purlins and wall plates included) and not for tie-beams. The assembly modes that we reproduce in figure 13 imply the use of a keystone (in black in the fig.) set in by force, which ensures excellent sturdiness. These assembly forms are extremely varied. Some reproduce "the line of Jupiter" assembly used by European carpenters, not always well assimilated in these cases.

Despite the development of masonry, wood is still the most commonly used building material in most of southern Asia. However, due to its fragility, there are very few clues to what prehistoric wood architecture looked like. The only elements able to give us any ideas are the embedding of beams, but this is rarely seen.

Even when the base of the building is in stone or brick, the roofing rests on a wooden structure, the framework. This was the case as early as the prehistoric period. When it was possible to reconstitute this roofing, we noticed that it consisted of a simple flat roof set on beams parallel to the desired pitch of the roof. Due to its location at the edge of the hill, the remains of one of the houses of level III/1[33] at Mundigak were high enough so that the embedding of the beams was preserved (fig. 14). This is a small house measuring 3.75 x 2.60 meters inside with very thick walls (0.60 m) made of laterite mud. The house is lit by a window that resembles a murder hole, 0.20 m wide and 0.75 m high. A hearth was set up in the centre and it is probable that the inhabitants entered the house through the roof, despite the narrowness of the home. The sturdiness of this building allowed it to survive the disappearance of level III/1 and resulted in its remains being reused on level III/2. At that time, the house underwent a restoration. Two buttresses were added against the longitudinal walls and the roof was probably redone at the same time. The roofing was supported by four beams set in the side walls. As this was a restoration, the embedding had a complex form that made it possible to lay the beams even though the masonry was already finished. These beams (we did not see any casements for lateral beams) would have been set directly under a woven reed bat supporting a layer of laterite mud. Debris of this mud showing the imprint of the supporting bat was discovered during the dig. This type of roofing has been used in Afghanistan, Pakistan and India through modern times. In India, to

[33] Jean-Marie Casal, *Fouilles de Mundigak*, Paris 1961. Vol. I, pp. 38-39, fig. C and Vol. II, fig. 13 and 14.

cover larger spaces, builders increased the number of supports and crossed the beams to create partitioned ceilings which are sometimes represented. An example is the roofing of cave I of Ajanta which dates from the 7th century[34] (fig. 15). Due to the vast surface area covered, about 400m^2, the ceiling is composed over three planes of beams. This would only be possible with a certain number of vertical supports, which we reproduced from painted corbels. Goloubew had already indicated for the central part of the ceiling[35]: "The rectangular frame where the large middle wheel window was inscribed is made up of four oblong and four square panels between which narrow parallel strips, with or without ornaments, are fitted. These strips most likely simulate the corbels (seen from beneath) of a support, the latter being represented (in cross-section) by the square panel set in the four corners of the frame." For our reproduction, we simply applied this principle to the entire ceiling, which proved difficult. On a square plane, there is a row of twenty supports set out in the rock (they are indicated in black in our figure). They are situated 3.50 metres from the walling, which leaves over 280 m^2 to cover. Thus, in addition to the pillars marking the boundaries of the space reserved for the dome, we reproduced two additional concentric rows, taking into consideration the location of the painted corbels (the base of these pillars is hatched in our figure). The form of this large room with a ceiling was probably highly in use. Its appropriation into stone continued in southern India until the end of the Cola period. In the west, it was used for the Hampi monuments with a different layout of the pillars on this site (the plane of the entire site is cruciform), but the principle remains the same[36]. Nevertheless, as early as the construction of the temple of Tanjore, this type of ceiling was constructed with a radiating framework, which was probably not the case for that in Ajanta. The date of the switch from a ceiling functioning as roofing to a ceiling marking out a roof that hides the framework can be placed sometime at the beginning of the 11th century. This type of roofing was most likely very widespread,

---

[34] Cave I of Ajanta was the subject of a very complete publication: Victor Goloubew, *Ajanta les peintures de la première grotte*, Ars asiatica X, G. Vanoest Paris & Brussels. 1927. The ceiling is featured in plates LVII to LXXI; the drawing of fig. 15 was made in reference to these diagrams.

[35] Goloubew op. cit. p. 42.

[36] For these ceilings, we refer to Pierre Sylvain Filliozat and Vasundhara Filliozat, *Hampi-Vijayanagar, the temple of Vithala*. Sitaram Bhartia Institute of Scientific Research; New Delhi, 1988. In particular, plates 20 and 29.

but it was not the only type. The ancient forms of wooden architecture and carpentry are attested to by the represented architecture, either painted in India as at Ajanta, or sculpted as in Mahabalipuram. This is also the case of numerous caves as we just saw, at Ajanta, caves 1 and 26, or at Karla, or even in Afghanistan, Bamyan[37] or Foladi[38]. These last two feature a stacking of beams, a technique used in the roofing of prehistoric terraces with a space reserved to let the smoke out. This technique has lasted until modern times in Badakshan. However, it seems to have completely disappeared in southern Asia, except in Kerala, which has proved to be a real conservatory of the ancient carpentry of southern Asia. It is also possible that even when the construction was accessed through a door, the space reserved in the centre was a representation of the ancient roof access in this region, just as we have seen in Mundigak. This was still in use at the beginning of the 20th century in the centre of Afghanistan, for example in the village of Kupruk near Band-i-Amir. In this village, winter is severe. Heavy snowfall can reach two to three meters in depth, almost burying the houses completely. Hence, the need for access by the roof. Nevertheless, due to the placement of beams set in the corners, this type of framework made it possible to cover a larger space. In Foladi, the beam system of cave C can be entirely reproduced (fig. 16). The plane is square but only comprises three walls. The missing wall is replaced by a large opening that prolonged the room. It was covered on the outside by a canopy whose framework was made up of beams left overhanging from the upper level of the framework of the room. The lower level is formed by beams set in the corners. The entire structure was pegged together with sturdy plugs, also shown. The central area, probably representing the space left free, is sculpted in the form of a small dome set on a circular fluted moulding.

Cave XV in Bamyan is remarkable. Not only is the ceiling sculpted like those of Foladi, but it is also set on three planes (fig. 17). What truly distinguishes this cave is the walls, which are also sculpted. They represent a truncated pyramidal wooden structure (fig. 18), most likely on a reduced scale. The square apex of the void, at the level of the

[37] A. and Y. Godard and J. Hackin, *Les antiquités bouddhiques de Bâmyân*, Paris 1923, and J. Hackin and J. Carl, *Nouvelles recherches archéologiques à Bâmyân*, Paris 1933.

[38] B. Dagens, M. Le Berre and D. Schlumberger, *Monuments Préislamiques d'Afghanistan*, Paris 1964. The study of the cave monastery of Foladi by B. Dagens forms the second part of this work.

wall plates, actually measures only 2.02 meters on the sides. This is quite a small surface area for the volume of wood represented.

It is certain that the builders wanted to cover a structure of increasing proportions, which would no longer allow for such an extremely heavy system. In Badakshan, the largest houses observed are no wider than 6 meters on each side, with significantly wide beams measuring 40 cm on each side. In the Trivandrum region of Kerala, India, this type of carpentry was used up until modern times, with one major difference. The carpentry was in fact the décor of a large ceiling supporting roofing with four sloping sides, which led to the placement of a small dome in the centre, similar to the one on the sculpted framework at Foladi. To obtain a covered surface area of about 45m², the builders set up five levels of beams. This shows the progress of frameworks in India. Probably due to the fact that the slant of flat roofs does not allow for a rapid evacuation of rainwater[39], the builders increasingly slanted the roof and then doubled it in two sloping sides. This led to the use of a support with a "bending" framework form. At about the same time in India for the same reasons, builders began to use curved roofing set on screeds. In Southeast Asia, the initial point is also a flat roof as referred to in the legends based around the appearance in Vietnam of "the mysterious virgin of the ancient heavens"[40]. She stands before her students with her arms akimbo and explains to them that this is how they must build their houses from now on. This led to the construction of the frameworks noted by L. Cadière, as well as another type of "taut ridge" frame.

As stated in our introduction, the "bending" framework was constructed according to a practical observation of the resistance of wood. When a beam is supported by two panel points and breaks under a given pressure, it does so not near its support but at a point about one-fifth of its length. This is what we call the shearing point. Hence, the builders tried to move the weight between the support and the shearing point as best they could in order to reduce the constraints at

---

[39] The rapid evacuation of rainwater is a vital necessity for buildings made of clay or even sturdier materials, but put together with weak mortar.

[40] L. Cadière, *Coutumes populaires de la vallée du Ngon-son*. BEFEO II, Hanoi, 1902, p. 373: "One day the 'mysterious virgin of the ancient heavens', standing before her students, placed her hands on her hips and showed them how to build their homes from then on. The figure of a man standing with arms akimbo actually represents the ancient form of Annamite houses, called *nha rôh* or *nha chi-Dinh* (house in the form of the character *dinh*)."

the centre of the beam and thus improve the solidity of the structure. This system, probably because of its simplicity, became widespread, not only in all of Southeast Asia, but also in most Chinese countries. This type of framework (fig. 19) features several tie-beams separated by vertical pieces between them with an indispensable crown post, as short as possible, which bore the ridge beam. However, there are many exceptions in Cambodia. If the space to be covered was too small for a porch for example, the purlins were set into gables without intermediary trusses (Ph 8). In Ajanta, in cave XVII dating from the 6th century, an unidentified *jataka* is illustrated on a wall painting[41]. We can see a bending frame with two levels where the crown post passes through the upper tie-beam and rests on the lower tie-beam of the largest section. In northern Thailand, this form of bending frame was used up until modern times, sometimes featuring more tie-beams as numerous as the supports are spread apart. The bending frame is still widely used in Cambodia and Laos. In Insulinde, it is still used, but often combined with the upper triangulated part above a horizontal supporting timber holding up the hip rafters.

Apart from the Indian treatises, there are several texts from Southeast Asia starting from the 19th century describing wooden architecture in particular. These texts are mainly meant to set the tradition, whereas the Western systems, and in particular the Truss, had a radical effect on the roofs of buildings. In the monastery of the Salé region in Burma, there is a manuscript dating from 1869 that gives instructions, and notably geomancy, for the construction of the façades of religious buildings. The text is illustrated with beautiful drawings[42]. In Cambodia, a master builder of the royal palace wrote a text in 1954 concerning traditional construction. This work was translated by Madeleine Giteau[43]. The technical instructions of this text are few and far between. The text is basically composed of ritual information. However, by way of an example, here are some excerpts from the text for the proportional rules used to divide a Khmer house: "Rules

---

[41] A reproduction of this painting is published in the work of Douglas Barret and Basil Gray, *La Peinture Indienne*. Editions Skira, Paris, 1963. p. 20.

[42] A page of this manuscript is reproduced in J. Dumarçay, *The house in south-east Asia*, Oxford University Press, Singapore, 1987. Pl. 3. This text is conserved in the Lay Tha Kyaung monastery in Salé, and is called a *parabaike*, a "white manuscript", in comparison with other manuscripts that are written in white on black paper.

[43] Madeleine Giteau, *Un court traité d'architecture cambodgienne moderne*, Arts Asiatiques XXIV, Paris, Adrien Maisonneuve 1971; pp. 103-106; Pl. I to VII and fig. 1 to 7.

divide the dimensions of the house. Whether the house is at eleven, twelve, thirteen, fourteen or fifteen dimensions, we divide these dimensions into parts. For the dimensions of a Khmer house, we measure the width to determine one part. The length measures two parts and the height two parts..." This example clearly shows that what could be rather technical is vague and must, for the author, have been simply understood, common knowledge among all builders. It was thus unnecessary to write it down.

While maintaining, more or less simply, the same principle of moving the weight between the shearing point and the panel point, builders used many variations of this model. For example, in Thailand where bending frames were used almost exclusively, pagodas were often composed of a nave (Ph. 9) flanked on each side by two side naves covered with half-roofs with tie-beams that are connected to pillars holding up the roofing of the nave. In general, there are no variations. However, in the Chiang Mai region, the builders added a vertical piece parallel to the pillar (fig. 20) to better distribute the weight. In the Chiang Rai province, the pagodas were sometimes constructed on a cruciform plane, which called for a new form of framework set into the walls. For example, in Chom Thong, the Wat Phra That pagoda is made of a central nave flanked on both sides by side naves over which are attached covered porches with a bending frame. The connection between the two elements is made by collar beams blocked at the base in a tie-beam of the porch and at the top in the wall plate of the central nave (fig. 21). In other Hindu states of Southeast Asia, the bending frame is used. In Burma, it can be seen in Pagan with the purlins attached to the gables of several monuments (Ph. 10). The Burmese master builders of the 19th century sometimes set the ceilings of the largest halls at the height of the tenth tie-beam. In Amarapura (at the Pyi monastery) for example, the visible part of the rafters is panelled. This is also the case in Salé in Burma where the main hall of the pagoda of the lacquered Buddha (fig. 22) has a square base and is covered with a visible bending frame. All of the rafters are panelled up to the level of the upper tie-beams, topped by a square panel measuring about 1.50 meters on each side. This last example is probably an imitation of a radiating framework. In Burma as well, in Amarapura, the builders adapted the panelled bending frame to roofing, one part of which has four sloping sides. This is a variation of roofing with a break and with a small pediment (fig. 23). The interior space is generally cluttered by at least four trusses, as is the case of our example. This type of framework is probably linked to an ancient model that is reflected in the sandstone architecture at Preah Vihear, a major site of Khmer civilization (early 11th century) at what is now the border with Thailand. One of the buildings that makes up this sanctuary (hall N)[44] is set on a rectangular plane, divided into three naves by two rows of pillars with capitals that are topped with a vertical piece meant to support the trusses and wall plates of the central element (fig. 24). This left a space for sloped windows that illuminated the central nave, the side naves being covered by a roof break set over two trusses attached to the pillars. This also corresponds to the roofing of the hall, which is in front of the cella of the Vat Phu sanctuary. Here, the side naves are covered by overhanging roofing supported by small trusses composed of a stone strut on which a wooden ceiling strut is attached[45].

In the north as well as in the peninsular area of Thailand, probably due to Burmese influence, roofing with two lone breaks (under side gables) is widespread. When it wasn't possible to construct roofing with four sloping sides due to a lack of means, the builders set up false gables over a radiating or bending frame. This was done in almost all of southern Asia, India, Kerala, Burma, Thailand (fig. 25), Cambodia (in the Battambang region) and Laos.

In Laos, particularly in Luang Prabang, the builders adapted the bending frame to roofing with four sloping sides (fig. 26), which led to the use of hip rafters at the base with very diverse combinations.

Due to its success, the principle of the bending frame was not always properly followed, particularly in India where aberrant constructions are numerous and in every shape and size. The most common technique, used in Tamil Nadu, consists of moving all of the weight by putting several vertical ceiling struts on a single tie-beam. To maintain the spacing of these ceiling struts and give a consistent appearance to the structure, horizontal pieces that sometimes crossed the ceiling struts were added. In southern India, we can also see that the crown post was no longer used. Instead, the ridge rests on a curved beam, pushing the weight onto the ceiling struts (fig. 27 and Ph. 11). This is

---

[44] Preah Vihear is described in H. Parmentier, *Art Khmer Classique*, Paris 1930. The layout of hall N: Pl. LVII.

[45] A photograph of this architectural detail is published in my study *Charpentes et tuiles khmères*. Paris EFEO, 1973. Ph. 8.

the most coherent transformation of the technique. It is even a vast improvement of the original process as the weight of the ridge is no longer set over the centre of the tie-beam.

There is one major disadvantage with bending frames. They take up much of the interior space of the building. Hence, builders came up with a number of solutions that leave a larger space free while sufficiently ensuring the support of the roofing. It is probably for this reason that the Burmese set the ceiling at the level of the tenth tie-beam, as at that level there is enough space between the beams so that the space does not seem too cluttered. However, this is only a last resort.

In India, as early as before the 7th century, master builders used a radiating framework made up of a newel to which a part of the rafters is attached. The rafters are blocked at the base of a wall plate, which makes it possible to completely clear the covered space. This technique is not exclusive to southern Asia. It has been used for a very long time by the nomads in central Asia to cover their yurts. These nomads came across difficulties similar to those faced by the Indian master builders in terms of keeping in place the screeds attached to a single newel. As with the prehistoric architecture of the region (in Mundigak, for example), the yurt often encloses a central hearth whose smoke is extracted at the centre of the structure. Thus, the principal piece of the yurt is a ridge finial composed of a circle of wood (fig. 28) to which are attached the screeds that support the felt and leave room at the centre of the structure for smoke evacuation. One of the oldest examples of the radiating framework is shown in Draupadi Ratha in Mahabalipuram (Ph. 12). This technique is described in the *Mayamata*[46] treatise. When the building to be covered is on a circular plane, all the rafters attached to a single newel are evidently of the same length. This is not the case for square or rectangular planes. This difficulty raised the interest of several authors of treatises. B. Dagens reconstituted one of the solutions: the problem can be reduced to the calculation of the hypotenuse of a right-angled triangle of which the sides of the right angle are known. The method proposed by the text

[46] *Mayamata op. cit.* Ch. XVIII. The upper elements of the buildings § 30 and 31 and figures 19 and 20. I studied this text from an article published in 1973: J. Dumarçay, "Les charpentes rayonnantes sur plan barlong ou carré de l'Asie méridionale" BEFEO LX. Paris 1973, pp. 85-104. Pl. V-XII.

is not very rigorous, but this isn't very important as in practice, only four or eight rafters maximum reach the newel.

This technique was widely used in southern India. It is still in use in Kerala with variations, including the setting up of a "horizontal framing in the roof" which prevents the rafters from bending (fig. 29). The advantage of this technique is that its set-up allows for the benefit of a larger space. To ensure this, the "horizontal framing in the roof" is generally placed as high as possible, sometimes even encircling the newel. Radiating framework is also used in Insulinde, in particular in Java, Bali and Lombok. But in Bali, the builders still use the technique that entirely clears the interior space (Ph. 13). The difficulty in setting up the newel led to other forms that set the newel either on an axial pillar, as in Java (fig. 30), or on a crown post attached to a tie-beam (Ph. 14) as in Bali and Lombok. In Java, the radiating framework is used for one of the most characteristic structures of the Javanese civilisation: the *pendopo*. This is a large pavilion on a rectangular plane with a roof that is divided into three concentric parts. The upper part is set on at least four principal pillars that support a complex frame that serves as a wall plate for the ridge. A tie-beam hidden by a ceiling is set into this frame. In the centre of the tie-beam, a crown post supports a ridge to which are attached radiating rafters that are set at their base on the wall plate. At their ends, some of the rafters are crossed through with iron rods which support the frame of the second part of the roofing. There is a similar frame for the lower part (fig. 30). Today, due to European influence, frames are still used. The rafters however are no longer radiating, but perpendicular to the frame (Ph. 15).

As we have seen, radiating screeds have long been used by the nomads in central Afghanistan to support the felt of their yurts. In India, however, the use of screeds is proven by the numerous sculpted Buddhist caves. In Karla in the Poona district, for example, a large cave sanctuary from the 2nd century[47] is comprised of a large rectangular hall with a semi-circular apse sheltering a stupa. The nave seems to be covered by a vault roof structure set on screeds attached to a wall plate supported by pillars with no intermediate trusses. This implies that the

---

[47] Debala Mitra, *Buddhist Monuments*. Sahitya Samsad, Calcutta. 1971. pp. 155-156.

screeds are extremely robust, 40 x 40 cm at the smallest part of their diameter. In Mahabalipuram, one of the sculpted sanctuaries entirely set clear of the rock—the Bhima Ratha (7th century)—is covered by this type of roofing (Ph. 16). Many cave or constructed sanctuaries of the same era feature a structure that is covered by roofing supported by screeds and that is finished on one side by the hip roof and on the other side by a gable that breaks the line of the vault (Ph. 17). The gable is decorated with the representation of a truss supporting the screeds, sculpted into the prepared stone wall like at Tiruttani, for example (fig. 32). This makes it possible to reconstitute the previous framework, which differs from that reproduced in Karla or Bhaja in that it is comprised of at least one truss. This type of framework has almost entirely disappeared in India. However, it is still used for small sanctuaries in Kerala and combines the radiating framework on the slope of the roof and the screeds for the nave (Ph. 20). In Southeast Asia, this type of roofing had a deep, widespread impact, particularly in Cambodia where it seems that builders knew only of the represented form of the technique. A corbel covering a rectangular space of galleries is shaped with curved roofing that evokes the roofing over screeds in India. This entailed a great number of inconsistencies. For example, the roofing tiles represented are Roman tiles and cover strips which can only be laid on certain planes, never on curved ones. In their representations, master builders had an eye for detail. The butt-end tiles are carefully sculpted and already hint at the models of contemporary roofing tiles (see Ch. IV).

In Southeast Asia before the implementation of Hinduism, there were most likely many types of roofing before the Indian techniques became the norm. Two of these ancient forms have survived. One is a very simple roof with four sloping sides. Its framework is reduced to a tie-beam supporting a crown post, which itself supports a ridge set amid the four sections forming the roof (fig. 33). This structure does not exist in India, but is still greatly in use in Java, Bali and most of the Sonde islands. This model is shown in its simplest form in the cave monastery of Goa Gadja. The other form of framework dating further back than the implementation of Hinduism is the "taut ridge" frame, which is still sometimes used. This technique consists of moving the gables outwards to overhang, thus "tightening" the ridge that supports all the weight of the rafters without the need for an intermediate truss. This leaves a large space clear inside the structure. Under the weight of the rafters, the ridge curves inwards, giving the roofing its particular shape[48]. The first proof of structures covered using this technique was discovered in Yunnan, China (fig. 34) with a model represented on the cover of a bronze cowrie holder[49]. This tiny building rests on two pillars that support the two overhanging gables. Under the weight of the rafters, all reproduced in the model, the beam is slightly curved inwards, giving the building its characteristic appearance. It is probably this aspect that incited the legend mentioned earlier, as many Insulinde carpenters use as a mark an isosceles triangle whose height crosses its base, evoking the goddess with her fists akimbo.

This type of building appears on the projecting areas of Borobudur, as well as on those of Prambanan. On this temple, most of the buildings we see are covered in this way. It is likely that the 9th century is the period of the greatest expansion of this type of roofing. Many temples of Champa represent a wooden building covered like this. In Java, around the 13th century, the "taut ridge" framework took on a religious significance related to the worship of Vishnu. This is particularly the case in eastern Java. For example, the porch of the Pari *candi* seems to be covered with roofing set over a "taut ridge" frame. This is also the case of the doors of the Belahan sanctuary. The element that seems most characteristic of the religious significance of the technique is the numerous models of stone silos that represent a building of this form, which sometimes features one of the attributes of this god (most often the conch shell) sculpted in relief on the roofing. Due to the reduced availability of large vats as well as the difficulty of setting up the ridge (we see the same type of obstacles that prevented the increasing use of the radiating framework for the implementation of the newel), builders set up bending roof trusses inside the houses. This took up the interior space and thus negated the benefits of using this technique. The same results are seen when builders place a pillar under the newel of a radiating framework. But no matter what difficulties arose, the builders always made an effort to preserve the ancient appearance of the structures. For example, in Laos and in northern Thailand, builders constructed libraries with overhanging walls. This

---

48 J. Dumarçay, *La faîtière tendue (Histoire d'une technique)*, BEFEO LXX, Paris 1981.

49 M. Pirazzoli-t'Serstevens, *La civilisation du royaume de Dian à l'époque Han*. Paris 1974. Our figure is drawn after figure 14 of this work. A bronze mirror of the same period in Japan was recently discovered, featuring on its other side the silhouette of a house on piles with a "taut ridge" roof.

implies the use of the properties of the resistance to the traction of the wood, but builders abandoned this technique and preserved the appearance despite the use of "bending" trusses (fig. 37). In the 17th century, builders in southern Asia learned how to use the truss. The technique spread rapidly, without however changing the shape of the roofing (Ph. 19). Sometimes, in imitation of triangulation, builders added aberrant elements to the framework (Ph. 20).

If we attempt to describe the evolution of carpentry in southern Asia, even though it may seem elementary, it would seem that in India, the flat roof set over beams perpendicular to the supports was widely used. This method was followed by a multitude of techniques that were often used together. On a single relief at Amaravati (fig. 36), three carpentry techniques can be seen, including a radiating bending structure and a structure on screeds. This is also the case at Mahabalipuram. But soon after the 9th century, carpentry on screeds was no longer used for large buildings. It is in this smaller form that it appeared in Southeast Asia, where it is still being used today (Ph. 21).

In Burma, Thailand and Cambodia, only "bending" frameworks—in diverse forms—lasted until modern times. In Indonesia, varied forms have been preserved, but the ancient techniques have more or less disappeared. Builders no longer use a taut ridge except for silos, nor do they use radiating frameworks, except in Bali. Even bending frameworks are no longer used in a coherent way except on rare occasions, and particularly by Chinese carpenters who work on the archipelago (Ph. 22). It is the European truss that has truly imposed itself, although it has not prevented the use of other architectural forms that have been preserved.

CHAPTER THREE

WOOD, SUPPORTS AND STILES

Wood was used not only for carpentry, but also for bearing structures. In southern Asia, these took the form of pillars with a square, and more rarely, circular form.

In India, as early as the 6th century, pillars were salient, projecting from the surface of the wall. This specific feature was taken up by the Pallava master builders[50]. For example, the small temple of Panamalai (Ph. 23), constructed in the beginning of the 8th century, is composed of a structural body with a cruciform shape. Its corners are marked with pillars set on mythical animals, lions or goats rearing up on their hind legs, projecting from the external side of the wall. This demonstrates the sumptuous layout of the slightest modern and probably contemporary rural home of the Pallavas. In this structure, the supports of the framework project from the external side of the wall. They are very often made of panels of woven bamboo strips, set over the inner side of the pillars. This style can be seen on many houses in southern Asia. However, due to the generalisation of the use of the saw, it became easier to obtain planks of wood. Thus, wooden partitions were set up, particularly in Thailand (Ph. 24) and to a lesser extent throughout all of the countries of Southeast Asia.

Vertical supports are used in a simple manner. They are either set directly on the ground when the home is set on the same level, or they are used as piles with an elevated floor.

The staking of the pillars entails driving the pillar or pile forcefully into the ground or hole made beforehand, and closing up the hole after the pile is set into place. The supports can be set over a preparation or a base of stone placed on the ground. It goes without saying that there are many variations of this, depending on the location.

The architecture represented on the monuments displays many con-

---

[50] This set-up was most likely used in order to clear the base of the walls from the "projecting area", as the cornice followed the capital at the top of the pillar or simply the wall plate. In southern Asia, due to the monsoon, the "projecting area" often resulted in the laying bare of the walls, despite precautions.

structions that use piles. In Borobudur[51], the projecting features show that sometimes the pillars were not only raised from the structure (fig. 37), but that they could also be completely set apart from them. On a relief of the first gallery on the northern side, there is a representation of a covered silo with a "taut ridge" (fig. 38/a). This small structure is set over piles driven directly into the ground, which support the floor. Just under the floor beams, they are topped with a sturdy circular structure, which is meant to prevent rodents from entering the silo. This particular structure is still used in all of Insulinde, no matter what construction technique is used. On a relief on the western side of the same gallery (fig. 38/b), there is a representation of a building shown with the layout of pillars. Their base, shaped in a hemispherical form, is set into a stone supported by the ground. Its upper section is indented to support the semi-round base of the pillars. This form, no longer used today, allowed for an excellent adjustment of the layout, but called for the perfect placement of the stone bases. In Laos, where the pillars were usually set into stones, the adjustment is not made at this level. After the pillars were placed upright and before the wall plates were put into place, the builders adjusted the pillars depending on the layout imposed by the base stones[52].

In Angkor, the dig led by B.P. Groslier, which preceded the reconstruction of the terrace of the Leper King, made it possible to recreate the base of the pillars of the pavilion supported by the terrace (early 13th century). This base comprised a cross made of two sturdy planks inset at half their length. The pillar was put into place over this intersection. Then, the filling was done between the support walls. This process of driving the base of the pillars into the ground is still used in Cambodia and Laos, particularly at the edges of roads and paths in areas liable to flooding.

In much of southern Asia, pillars were often driven by force into the ground until the builders seemed to have reached a more solid layer of earth. This process is very popular in Burma, where it caused the ruin of these constructions. The evaluation of the ground resistance was

---

[51] Following H. Parmentier, we studied in detail the architecture represented on the salient structures of Borobudur in our *Histoire architecturale du Borobudur*, Paris EFEO, 1977, Ch. I, "le milieu architectural".

[52] Much of the information on Laotian homes I owe to Sophie Clément-Charpentier and Pierre Clément, *L'habitation Lao*, Editions Peeters, Paris, 1990, particularly Chapter XI.

often mediocre, and after a certain time, the pillar begins to sink (Ph. 25). This is also the reason for the increased number of panel points (Ph. 26). This technique was also developed for reasons of simplification. Thus, in Minangkabau regions, when the ridge was no longer taut, it—as well as the other purlins—was set directly over the pillars, which led to the lack of space in the principal hall (fig. 39).

At their top, the pillars directly supported the longitudinal beams of the roofing. However, the bearing surface is often enlarged by inserting a double corbel between the support and the beam, which was done in Ajanta as early as the 2nd century, and in the sculpted Pallava caves in the 7th century. The addition of corbels was rarely seen in Southeast Asia. To support a heavier load on a single pillar, builders used brackets, for example, to maintain the forward thrust of a canopy or simply the projecting section of the roofing. This technique was adopted by the Khmers. Traces of the use of wooden brackets in the stone pillars of covered galleries are quite numerous, for example in the surrounding gallery of Banten Samre (12th century).

Exceptionally, the supports are diagonal instead of vertical. This is the case of the bridge represented on a relief of the fourth gallery of Borobudur; this technique is still used today (fig. 40). To cross the bed of a small river, four pieces of wood—usually very sturdy bamboo—are stuck diagonally two by two on each bank. They meet at their points, forming the sides of the bridge which holds up the bridge deck.

In most rural habitations of the region, the walls were made of mats of bamboo strips measuring 2-6 cm in width, and using different weaving styles. These mats are placed on the inner side of the pillars that can be seen outside the building. In most cases, the windows are constructed without woodwork frames. They are simply rectangular openings cut into the mats. A small panel of woven bamboo serves to shut the window. It is attached on one side to the top of the outside frame of the window and held open by a "tie-beam piece" block[53] that was simply removed in order to shut the window.

Once the use of the saw became more common, the walls of the richest homes were constructed with planks, either attached horizontally on the outer side of the pillars, or vertically on the horizontal beams that united the pillars inside the building. These same beams

---

[53] We use this term to describe a small wooden block, by analogy with the shape of the small piece of carpentry of the same name which is sometimes used to connect the principal rafter to the wall plate.

provided the base for the windows, except in the Mandalay region of Burma where builders continued to leave the supporting structure visible outside the building. In Cambodia, walls made of vertical planks were reserved for only the most important buildings, for example the guest lodge of a monastery. The windows (Ph. 27) are composed of an external frame, which is often quite elaborate, and two shutters that open outwards. The Khmer stone monument architecture largely features a type of window (fig. 41) that is no longer used in modern Cambodia, but which is still used in Thailand. It is a quadrangular frame constructed on the plane of its diagonals, which led to inset features reproduced in the stone. As early as the beginning of the 9th century, master builders soon realised that this type of embedding considerably weakened the stone, and they only used this model as décor on the façade. Turned bars are set inside this frame. This is actually a transposition. In stone, it is not possible to turn elaborate bars of such small width like those turned in wood, but the desired outcome is the same: letting the light enter the building while ensuring a closing method. This transposition shows us to what extent the location of the windows was chosen with care. For example, builders avoided setting the trusses over the site of the window lintels, as demonstrated by the placement of the tie-beam casings in the inner cornice of room N of Preah Vihear (fig. 42). From the end of the 12th century, several Khmer monuments featured sculpted false windows, on which a half-lowered shade was represented over half-bars (fig. 43. This window demonstrates a point: the bottom bar of the shade became blocked when it was lifted and the cords slack). Contemporary master builders no longer use this method. In Java, this process of window closures with turned bars is well represented on the relief of Borobudur, but it has not been transposed into stone architecture.

In Thailand, Malaysia (fig. 44) and Java, the inner walls often featured beautiful woodwork. They were made of panels made of ribbed frames fitted with planks. In Java, for inner walls as well constructed as this, the integration of windows was also carefully conducted. In the old houses of Kudus, the lintel of the opening frame often has a half-circle shape. This complication was taken up in the design of the inner walls of the house (Ph. 28). In Insulinde, when the windows are closed with bars, these bars are not turned, but instead have a square or octagonal shape.

While in prehistoric periods houses without windows were relatively common, there was obviously always an entryway. As early as the third millennium in the northern civilisations of India, these entryways were closed from within with a door whose axis turned in a socket of the door pivot at the base. At Mohenjo-Daro, the major site of the civilisation of the Indus, the doorways were narrow—about 0.90 metres—and the door leaf was perhaps set into a frame[54], considering the small number of sockets of door pivots discovered. These were cut into baked bricks, and more rarely, into stone. On a contemporary rural site of Mohenjo-Daro—Amri—the excavators discovered several stone sockets of door pivots but none of them in place. This is also the case on the historical site of Taxila, despite the large number of houses unearthed (probably only their foundations). The Buddhist monasteries that have been excavated around Taxila and whose remains are above the level of the doorways show no indications relating to their stiles. We can make the same remark about the monasteries encircling Sanchi. Yet, on the raised surfaces decorating the porticos of this site, a great amount of architecture is represented. For example, on the western pillar of the north door, the raised surface represents the departure of the king of Kosala leaving Sravasti, his capital. He passes through a monumental door whose architectural detail is represented with great care, yet which does not show any indication of the stiles[55]. The large stupas of Sanchi are encircled by a balustrade, intersected at its axes by porticos with no traces of closing elements. At Bharhut, on the monuments represented on the raised surfaces, no stiles are observed. However, one relief shows a door with a tied curtain[56]. This may have been more common than this single representation would lead us to believe.

On Pallava monuments cut into the rock, if there were doors—which we cannot be sure of—they must have comprised a door frame resting on a doorway cut out of the rock by the sculptors. On monuments constructed in the same period, in Panamalai for example, the doors were placed in a frame whose casing was set apart in the masonry. On the Shore Temple in Mahabalipuram, the door of the L sanctuary

---

[54] This is the hypothesis of H. Mackay, considering the small number of sockets of door pivots discovered on the site. E.J.H. Mackay, *Further Excavation at Mohenjo-Daro*. 1938, reprint 1976. Vol. I, p. 167.

[55] Sir John Marshall and Alfred Foucher, *The Monuments of Sanchi*, reprint of 1982. Vol. II, Pl. XXXIV.

[56] Alexander Cunningham, *The stupa of Bharhut*, reprint of 1962, with an introduction by Da Vasudeva Agarwala. Pl. XXX/1.

was also set in a frame located in such a way that when the door was open, the two door leaves did not encroach upon the cella. This layout, which called for walls of a certain thickness (at least equal to the width of the leaves), was not the common standard throughout the site. The H sanctuary does not use this method and when the door is open, the door leaves enter the cella. At Tiruttani, the frame was held back to a doorway in which sockets of door pivots were probably placed. The top of the axis of the door leaves was set into pivot stones whose casings were set into the stone lintel of the door. This latter form lasted in Southeast Asia, for example on the large temple of Angkor Wat, but was rarely used in India. During the Cola period, the doors of the large temples were constructed in frames, which were themselves set into casings laid out in the masonry. The common practice was to use door leaves that did not encroach upon the space when open, at least until the end of the 13th century.

Since prehistory, door leaves have featured the axis on which they turn. If the door had only one leaf, it would stop on the doorsill. When the door had two leaves, the leaves would be set one over the other on a rebate set in their width. This system has been used at least since the 8th century in Java. The closed doors represented at Borobudur hint at this process as well (as there is no banging represented, fig. 37). Nonetheless, on the temple of Candi Sewu (where over 250 doors were added on, following various methods which are only known to us through the casings set out for the placement of the sockets of door pivots[57]), the pivot stones consisted of a simple cylindrical casing inset in the lintel, no matter what model was used for the base. Due to the rigidity of the structure, the axis of the leaf had to be set at an angle. In Cambodia in the same period, the doors were made with closing stiles. A piece of wood at least ten centimetres wide was nailed over one of the leaves, usually the left one. The edge of the other leaf was pressed against it when the door was closed. This element is represented on the false doors cut into the stone (fig. 45) and was often decorated with square cabochons whose chiselling was similar to that of the door leaves. Depending on the quality of the building and the weight of the door leaf, the base of the axis that turns in the socket of the door pivot was often covered in bronze (fig. 46 a). The same technique was used

[57] I analysed these techniques in my study *Candi Sewu et l'architecture bouddhique du centre de Java*. EFEO. Paris 1981, p. 22, figs. 17 and 18.

for the pivot stone (fig. 46 b). When the door was open, it tended to pull on the pivot stone. After repeated use, this destabilised the lintel. To prevent this inconvenience, the builders used supports attached to the entry walls. This construction detail was common in Java as early as the 9th century, but was rarely seen elsewhere.

On the Pagan site in Burma, the archaeologist Duroiselle discovered 11th century remains of teak wood stiles[58] in the temple of Shwe-Zigon. There was a door with two leaves and a closing stile which were richly decorated, but these fragments do not give us any indications as to how the door was opened or closed. On this same site, the doors that separated the porches of the temple of Ananda[59] from the first circumambulation hallway and the central hallways leading to the cella are still in place. These doors probably date to the time of construction of the temple finished in the last years of the 11th century. These teak doors do not have stile frames, but do feature two leaves and a closing stile. The axis rests at the base in a socket of the door pivot set in the doorsill and turns at the top in a ring fixed into the masonry (fig. 47). The door does not really close off the space, as the leaves are latticed, and due to the absence of a frame, the space is clear between the top of the door and the vault covering the space. These doors open inwards and swing completely back against the inner wall. The leaf, although mainly hollowed out, is quite heavy. To prevent it from putting too much strain on its axis, a masonry support was set against the wall (see design of fig. 47). This closing principle was used for a long time in Burma. It can still be seen on the 18th century frescos of the small sanctuary of Ananda Okkaung in Pagan[60] (fig. 48) and on the raised surfaces that decorate the cornice of the Salé sanctuary (Ph. 29).

This layout does not allow for the conservation of traces of stiles on brick or stone monuments. Even on constructions that aimed to faithfully reproduce a wooden building such as the Swe-dan-gu[61] in Pagan, there are no traces of the insertion of stiles. In Mandalay, in the great wooden palaces of Kin-Kun-Mingyi for example, the stiles were made based on European models that were significantly modified.

---

[58] Gordon H. Luce, *Old Burma-Early Pagan*. J.J. Augustin Publisher, Locust Valley, New York. 1969-1970. p. 267 and Pl. 178-182. P. Pichard, *Inventory of monuments at Pagan*, Kiscadale, EFEO, Unesco. 1992, Vol. I, monument 1, pp. 61-67.

[59] Luce, op. cit., p. 274. Pl. 271-275. P. Pichard, *op. cit.* 2001. Vol. VIII, monument 2171, pp. 144-153.

[60] P. Pichard, op. cit., Vol. VIII, monument 2162, pp. 130-135.

[61] P. Pichard, op. cit., Vol. I, monument 73, pp. 144-145.

If the door was opened inwards, the builders maintained the closing stile of the Burmese model.

During the medieval period on the island of Ceylon[62], great care was taken with the finishing of the doors. The sockets of the door pivots and pivot stones were placed outside of the stile frame (fig. 49). Sometimes, the closing of the door was made possible by a moulding placed in the thickness of a door leaf that fit into the gorge of the other door leaf. This system allowed for an excellent closing of the door. However, the two leaves must be shut together so that the moulding is sufficiently caught in the groove (which is why each leaf had a ring on it). The stile frame is set over the masonry and often had a shape different from that of the leaves. From the interior, the door was closed by a double lock.

While in India the doors generally featured two leaves, in Ceylon they often featured only one. As a consequence, these doors were very narrow. The door that we have drawn in fig. 50 is only 65 cm wide.

In Laos, the doors are generally made of woven bamboo set on a frame whose jamb is attached to horizontal beams that hold up the partition[63]. There are also, however, very finely worked doors[64] (fig. 51) with two leaves and a large exterior closing stile so that the door can open outwards. But, especially for the inner doors, the closing stile was placed on the inside, and in this case, the lock overlaps it[65].

In Insulinde, builders continued to use doors with two leaves and without a closing stile. These doors had simple rebates set in the thickness of the leaves and were locked on the outside. However, in the centre of Java, for rectangular houses, builders put into place sliding doors, which pushed the entryways to the building near the corners. Even for large, carefully designed homes, the entryways were not set at the middle of the building as was the common practice (Ph. 30 and 31). In Bali (fig. 52) and Lombok today, the doors are all made with a stile that holds the sockets of the door pivots, some reduced to a

---

[62] Ananda K. Coomaraswamy, *Mediaeval Sinhalese Art*, Pantheon Books, second edition, 1956, p. 133; and Godakumbura, *Sinhalese Doorways*, Archaeological department, Colombo, 1970.

[63] See Clément, op. cit., P. 306 and fig. 284.

[64] H. Parmentier, *L'art du Laos*. Edition updated by Madeleine Giteau. EFEO, 1988. Pl. XLIX, wooden door of That Ing-Rang.

[65] For example, on one of the houses studied by the Cléments, op. cit., fig. 21.

very simple copper plate. The pivot stones are set into the upper bar of the frame with no particular placement. In Sumatra, the doors of Minangkabau houses are very elaborate and often preceded by a large flight of steps with stairs that make it possible to climb above the piles. These doors have two leaves that fold back against the inner walls. However, as houses were used less to house large families and more to house single couples and their children, the houses kept their general shape but on a smaller scale. In this case, the door was moved to the end of the principal façade.

Houses on piles did not always have doors as their entryways. Amongst the many representations of buildings on piles of the inner gallery of Bayon, there are very few staircases. It is likely that people entered the house directly through a trap door on the floor of the building. This technique is still common today in the Batak region on the island of Sumatra, and in the Toraja region of the island of Sulawesi. The inhabitants enter the home through a trap door placed in the floor of the house between the piles. Thus, the façades of the houses appear closed. Around the Toba lake in the Batak region, the models of houses (since 1920[66]) often encircle very elaborate tombs. There are no doors; this is perhaps a way to highlight the prosperity of the owners through the great detail given to the décor, which is not always evident. It could also be a way to prevent the dead from returning to the homes. The trap door is usually closed by a movable panel attached to the opening. It is blocked inside by a weight. The addition of a hinge appeared only recently.

Like other architectural elements, it is obvious that doors have undergone repairs and changes that we can only estimate from the rare remains. Nonetheless, the northern monumental door of Ta Prohm of Angkor was redone at a date that we cannot precisely pinpoint. At this time, due to their reinforcement, the new leaves no longer remained hidden when the door was completely open. The builders set casings in the jambs to attach the stabilising elements while the door was being opened (fig. 53).

---

66 See Daniel Perret, *Tombes batak modernes de la région de Barus*, in Claude Guillot, *Histoire de Barus, Le site de Lobu tua.* Cahier Archipel, 1998. P. 222.

CHAPTER FOUR

MUD AS A CONSTRUCTION MATERIAL

In India, mud has been used as a construction material since prehistoric times. First, unprepared mud was used to build dikes throughout Southeast Asia, and particularly in Cambodia for the edification of the "baray", the large reservoirs on the Angkor site. These dikes reached 12 to 15 meters in height and up to 60 metres in width (including the spreading caused by the deterioration of the dike).

In architecture, this material was used in the form of pisé, a mixture of clay and straw compressed in a formwork. Over this base measuring about 40 centimetres in height, the builders lay more rows of the same height until the top of the structure is reached (which is limited in height). Nonetheless, the 13th century remains of Lashkari Bazar[67] in Afghanistan—the pisé residence of Ghaznevid and Ghorid sultans—still reach a height of 14 to 15 metres today, with a base that is several metres in depth.

Pisé is not supple and requires high scaffolding for the set-up of the formwork. This did not pose a problem when the houses were low. For example, the house of Mundigak, whose purlin casements we observed (fig. 14), has an interior height under the ceiling of 1.30 metres. We may assume that as early as the third millennium B.C. when taller houses began to be built (in Mundigak, at the end of Period III)[68], the builders continued to use pisé[69] and began to use bricks. These bricks were made in different sizes and measured on average 45 x 22.5 x 10

[67] Bust, the citadel next to Lashkari Bazar, is called the door of India by Hudud al-'Alam (late 10th century).

[68] J.M. Casal. *Fouilles de Mundigak*, Klincksieck, Paris, 1961. p. 37. The use of unburnt bricks is exceptional during this period, but it becomes a common practice during the following period, for example for the construction of ramparts and palaces.

[69] In his definition of pisé, Littré remarks: "they stand joined slats together in two parallel rows between which they throw the mud which they then step on. When the mud is packed tight and slightly dried, the slats are removed". His definition of cob: "mortar composed of clay and cut straw, used for certain constructions. It is used to fill the empty holes, which can be quite large, on framework buildings". Despite the reference to cut straw, we use the term "pisé", which seems better adapted to the constructions of southern Asia.

cm. They were composed of clay which had been cleared of stones and diverse debris beforehand. In India, this clearing of the mud was often done by levigation[70]. The mud is set out to dry in the same state in which it was extracted, in small clumps—which are then ground into a powder. During this part of the process, all the stones and plant roots are removed. The resulting powder is watered down thoroughly and then left to settle. Often, a degreasing agent of sand, straw or rice hulls[71] is added to this mixture. The mixture is then carefully stirred together. For the highest quality bricks, the mud is beaten to remove any air bubbles. It is left to settle, and then the mixture is ready to use. It is placed in a mould, a simple frame dusted with dry soil for easy, and often instant, removal from the mould. The brick is then left to dry in the sun and turned over for even consistency before being used in construction. The pouring into the mould differs depending on the size of these bricks. For small sizes, the mould is filled simply, using no particular method. For large sizes, blocks of mud with the same volume as the mould but with a different shape are prepared. These blocks are significantly longer than the mould. The two sides that pass over the sides of the mould are folded in, which ensures better coherence as the folded area guarantees the presence of rice hulls at the edges. (When the mixture is stirred together, the added plant components tend to concentrate in the middle of the block).

There are many techniques for using unburnt bricks. In its early uses, in Mundigak for example, bricks were used for the walls and set out in two parallel rows. Sometimes the brick assembly showed greater detail, with alternating brick faces and headers. On the constructions of this site, bricks were combined with wooden elements: stiles and door lintels (the simple door leaf turned in a stone socket of the door pivot). The windows, on the other hand, were simple and narrow slits. A single brick formed the lintel and the shutter often consisted of a brick placed vertically in the hole[72]. The use of unburnt bricks was widespread in all of northern India and prevailed until the 16th

---

[70] This term is generally used in the pharmaceutical field for products that are reduced into a powder before being incorporated into a medicine.

[71] In some cases, when the rice does not reach maturity during very dry seasons, rice hulls are not used. Instead, rice powder and the rice envelope—the glume—are added, giving the mixture a particular appearance and in principle, better coherence.

[72] The window slits of the Mundigak houses start on the ground outside on which the "shutter" is set.

century. The line of small forts constructed on the north bank of the Indus during the reign of the Moguls is, for example, made of unburnt bricks. The external walls were very thick, sometimes up to three metres. However, for prestigious buildings, builders corrected the rough appearance of unburnt brick walls by using veneering. One example is the Kushan temple of Sukh Kotal, constructed with unburnt bricks. It has white limestone veneer and a sturdy interior wooden wall tie to prevent the structure from shifting, as mud buildings are very sensitive to climatic changes, humidity and extreme dry conditions which can cause the structures to shift.

For Muslim constructions of the 10th century, builders often used pisé and dried unburnt bricks together: two brick walls forming a double siding enclosing an area filled with pisé. This technique recalls another, which is extremely widespread in India for stone architecture, in which two sidings enclosed a filling of various stones bonded together with a mortar, usually mud.

From the time they were first used, unburnt bricks gave rise to a geometric décor. The first example that we know of can be seen at Mundigak. During Period IV at this site, in the third millennium, the main hill was occupied by a large mud building. The principal façade of this building featured a décor which we can only see on two levels. The lower level was composed of inset columns and the upper level of diamond shaped sections of bricks offset one over the other (fig. 54). This décor was covered with a plaster composed of clay freed of all impurities and with no added constituents. It was then painted white, except for the interior side of the doorjambs, which was painted red with a mixture of ground terracotta and tragacanth. This enhancement of the décor is one of the very first signs of an architectural focus in the region.

During the Kushan period, unburnt bricks were widely used. This often gave rise to very elaborate décor in which brick was sometimes used in different positions. For example, bricks were set diagonally to form the tip of the arrows on the décor of the arrow loops on the enclosure wall of Surkh Kotal (fig. 55) and the enclosure wall of Bactria. Muslim architecture gave this type of décor more grandeur and complexity. This décor survived due to better quality plaster, sometimes veritable stucco. It did not, however, lead to a particular technique of the placement of the bricks. The master builders of northern India most likely acquired skills in vault construction from Iranian influences. This is particularly the case in Lashkari Bazar where the builders used the barrel vault constructed over an arch centring or more often, in rows leaning against a drum (fig. 56). In some cases, very complex arch systems (fig. 57) were created. It would seem however that this was rarely used with mud as the building material.

Burnt Brick

Despite the abundance of mud constructions, whether made from pisé or sun-dried bricks, most of these types of buildings have disappeared except in northern India, due to climatic reasons. Unburnt bricks were gradually replaced with burnt bricks. The oldest remains of burnt brick buildings date back to the Indus civilisation. The procedure of making burnt brick is so simple that it did not lead to true craftsmanship before the historic period. Even today, in many regions of southern Asia, the bricks are fabricated at the construction site, which explains the inconsistencies in terms of the dimensions of the material.

Nevertheless, builders quickly realised that not all mud yielded the same quality brick, and that clay produced higher-quality elements. Thus, builders prepared very large temporary kilns near clay deposit areas (Ph. 32: this kiln was made in , west of Madras). There are many regions in southern Asia where the clay is suitable for making proper burnt bricks. So, whenever possible, the bricks were fabricated near the construction site.

Builders sometimes took advantage of other public works sites, for example the excavations necessary for irrigation, to fabricate bricks. We also observed a noteworthy practice (in March and April 1977) in the area surrounding the village of Sawanti near Magelang (in the centre of Java). The irrigation system of the rice fields there had to be modified by an average of 25 cm due to a general lowering of the water level. This left considerable excavated earth, allowing for the fabrication of burnt bricks. First, the level was lowered on the lowest rice field (fig. 58). This was done on the area to be modified, but only on ¾ of the area. The excavated earth was piled in the remaining area. After the stones were carefully removed, water was added to the lowered area. Then, the soil was mixed together and bricks were fabricated and set to dry on the area that had not been lowered and that had stayed above the water during the preparation of the mixture. Then, the kiln was put together, protected by a small shelter composed of some poles holding up matting. Secondly, water was added to the lowered part of the rice field again and bricks were fabricated once more and so on until the desired level was reached. Finally, the protective shelter was taken apart and fire stoked in the kiln. The resulting bricks were not of very high quality and the peasants sometimes found it difficult to sell them[73].

The implementation of burnt bricks has varied greatly since bricks first came into use during the development of the Indus civilisation. There is one technique in particular that was invented during this period to cover sewers. This corbel procedure was used all throughout southern Asia and is still used today without many modifications. When it was first employed, workers didn't take the procedure very far, never laying more than three or four levels of roofing bricks. However, there is one very remarkable exception: the drain of the bath of Mohenjo-Daro, which is comprised of seven corbel levels[74]. Burnt brick arches are a long-standing tradition. Among the many examples of this is the extraordinary vault of Ctesiphon, probably constructed in the 3rd century under the reign of Shapur I. The exceptional use of arches in this mud architecture reveals the knowledge of the technique, which is the same for burnt brick and rarely used in southern Asia. Despite the fact that there were numerous and large structures in Java and Cambodia, no arches were constructed there. In India, it is likely that burnt bricks were often used, even though the ancient remains seem few and far between. Nevertheless, from the few examples that have survived, we see that the voussoir arch was known and used up to the 10th century. There are examples in Mahabodhi and Paharpur. In Bengal[75], one of the oldest forms constructed copies the structure of a hut normally made of bamboo. Its translation into brick led to buildings that could only exist with a certain mastery of the semicircular

[73] The most interesting point to note in the fabrication process of these bricks is the use of a very long-term working plan (it takes several years to reach the definitive level over the entire area chosen for the operation). This involves the entire village. In the case of the operation at the village of Sawanti, whose population was less than 700 inhabitants, this process took over five years of work. Sometimes, operations of this type occur on a very large scale. This was the case of the area surrounding Candi Plaosan, where the entrepreneurs were able to fabricate good quality bricks.

[74] A photograph of this arch is published by Mr. Wheeler, *The Indus Civilisation*, second edition, 1960, p. 31 and Pl. X, fig. A. The location of this work is indicated in our fig. 2, under the letter K.

[75] These indications on brick architecture in Bengal owe much to the work *Brick Temples of Bengal* from the Archives of David McCutchion, edited by George Michell. Princeton University Press, 1983.

arch (fig. 59)[76]. This type of building with the same form of roofing was also used in Burma, for example in the Hsin Buy Shin group[77]. In southern India, the remains of brick structures are rare. However, the importance of those that are left points to a large use of burnt brick. A good example is the crown of the Amman temple from the 13th century in Darasuram, which shows a certain mastery of the use of the semicircular arch and the corbel. This allowed the master builder to move from the initial cruciform plane to the square plane of the crown (fig. 60 and 61)[78]. Despite its use in a Hindu temple, this structure was probably inspired by Muslim styles, copying the intrados of the vault of a building covered with a double dome. Brick vaults were also used to support the heavy towers forming the entry pavilions of certain temples, in Chindambaran for example.

In all of southern Asia, burnt brick was used most in Burma in very large buildings. One example is the temple of Ananda in Pagan, which is on a square plane and which measures 90 metres on each side with a maximum height of 40 metres. On the same site, the masonry of the large stupa of Mingalazedi[79] (fig. 62) is about 40,000 m^3 in volume, which represents approximately 7,500,000 bricks, which would have had to have been baked in no less than 150 kilns similar to the one shown in our Ph. 33. This type of kiln can only produce 50,000 bricks, with much waste left over. The volume of bricks mixed together in the 13th century on the sole site of Pagan is considerable, if not to say incalculable, and thus would have required rigorous organisation. For the baking, the fuel had to be collected, even if it was of poor quality (in the Indus valley, the workers often settled for reeds). The heat necessary for the baking was only achieved by draughts and a constant

---

[76] An axonometric drawing of the exterior of one of these 17th century buildings was published by Klaus Fischer in *Studies in Indian Temple Architecture*. Papers presented at a seminar held in Varanasi, 1967. Edited by Pramod Chandra; American Institute of Indian Studies. 1975. P. 184.

[77] For example, buildings 685, 686 and 687 of the inventory by P. Pichard, op. cit. Vol. 3, pp. 206-207.

[78] These two figures were drawn from plates XXII and XXVII by Françoise l'Hernault, *Darasuram, Mémoires archéologiques XVI de l'EFEO*, EFEO, Paris 1987.

[79] The temple of Mingalazedi is number 1439 on the inventory of Pierre Pichard, op. cit., vol. V, pp. 392-399. This monument was probably constructed in the 13th century. P. Pichard indicates two dimensions for the bricks: 48 x 23 x 8.5 and 38 x 16 x 4.5. No need to say that the number of bricks would differ considerably had the monument been constructed in a consistent manner using either one dimension or the other. Thus, we have chosen an average of 41 x 20.5 x 6.5 cm.

renewal of the fuel. In Pagan, the baking was very irregular, but even so, the bricks were used in the same condition in which they left the kiln. This explains why the mortar joints are very thick, sometimes up to 2 cm. The use of the bricks was extremely varied. The builders usually alternated the joints properly, which only rarely ended up directly one atop the other. To follow the line of the ornamental outline closely, the workers had to reshape the construction elements. In other cases, to prevent the outline of a string course from being adjusted with the bed, the workers had to place the bricks on edge as siding, wedged in at a rather difficult angle (fig. 63). They also used these beds of vertical bricks to help make the structure more stable. Moreover, bricks were sometimes used horizontally or on an inclined plane for the siding (Ph. 33).

The corbel was often used, not only for the roofing, but also as a string for stairways, which were rarely featured (fig. 64). What underlines the originality of the Burmese techniques is the use of the discharging arch, often at the base of a semicircular vault (Ph. 34 and fig. 65) when it forms the upper part of a light and when it is simple. It is often completed by a discharging arch for a lintel or a simple light in a wall. Sometimes, the discharging arch is doubled in the case of a semicircular vault, which acts as a buttress. The bricks that make up the arch are placed on edge. The central keystone is considerably reshaped, as are the other elements to a lesser extent. The discharging arch goes beyond the width of the light and adjusts the weight of the superstructures on the abutments, reinforcing the structure of the masonry without the intrados of the vault being weakened. This differs from Khmer and Javanese techniques in which the discharging cell of the vault (fig. 66), always as a corbel, has a width inferior to that of the light. This is perhaps a reminder of the constraint of the shearing point. It is not necessary to adjust the weight onto the abutments for the arch or corbel to be effective. The transition of the arches within the masonry of the structure is constructed with small brick elements that make it possible to regain the horizontal surface of the beds.

In Thailand, the monuments of Chaiya from the 8th century, located in the peninsular part of the country, feature a plane clearly inspired by Indian techniques. One may note that they share the same proportions as the Candi Kalasan in Java. However, from a technical point of view, they are built of brick set together with a lime mortar, like the Burmese temples. When the Khmer influence grew stronger, the master builders adopted not only the same shapes but also the

Khmer techniques for their brick monuments. This can be seen in Surin, in the north-eastern part of the country at the Sikhoraphum sanctuary dating from the early 12th century. The bricks are ground down one on top of the next and bonded together by a thin layer of binder composed of creeper sap mixed with palm sugar, giving the wall a smooth appearance and allowing for cornice overhangs. It was in Cambodia that true brick architecture began to be developed as early as the 7th century in Sambor Prei Kuk (Ph. 35). Many large brick structures were constructed here using the technique of grinding the bricks one atop the other, using water to obtain a smooth surface on which the binder is laid in an extremely fine layer. The grinding ensures perfect adherence: the smoother the surface, the better the contact. Brick buildings vary greatly in form, but technically they remained the same throughout the Angkor period. Even when stone began to be used for the large monuments, the corbels were still made of brick, as is the case at Ta Kev, for example (Ph. 36). Nevertheless, from the 12th century onwards, bricks were used less and less, without disappearing completely. Buddhist structures constructed during the 13th century only rarely used ground bricks and the joints became much thicker, sometimes over a centimetre thick.

While there used to be mud structures in Java, they have entirely disappeared today. However, burnt brick was used to build temples as early as the 8th century. These bricks were often square, as is the case of the enclosure of Candi Mendut. Although this form was never completely abandoned (the Candi Gunung Gangsir was built with square bricks in the 14th century), builders moved on to a rectangular model. The dimensions varied greatly, but we can observe that the ratio of length to width remains somewhat constant at about 1.65. These terms show us that the technique of implementation was modified and that the alternating faces and headers could not be regular, as on the Khmer monuments. To prevent the layering of joints, the bricks are all placed on their face sides and not aligned from one bed to the next, which is the case in Candi Jabung (Ph. 37). Apart from the protruding sections, there is no link between the siding and the inner base. This type of arrangement can be seen in Sumatra at the monuments of Padang Lawas (Ph. 38).

Roofing Tiles

Terra cotta is also used to make roofing tiles of all sorts for buildings. As we have seen, in the prehistoric period in India, the buildings were covered with a mixture of mud and straw placed on bats resting on the framework. This process is still used for many rural houses in the north, for homes made of *pisé* adobe. For wooden houses, foliage elements were often used. The oldest terra cotta tiles were flat, pierced at the apex with two holes in which two sorts of plugs held the tiles onto the lathwork. This model spread throughout Southeast Asia, particularly in Cambodia where it was used in Sambor Prei Kuk. However, this type of attachment technique was most likely abandoned at about the beginning of the 9th century and replaced by a catch, a small strip of earth placed under the tile connected to the lathwork, or by the folding back of the upper part of the body of the tile which then worked like a catch and hooked onto the lathwork. These two types of flat tiles are still widely used, with a number of variations. For example, on the island of Ceylon, builders made tiles in the form of a circular sector to cover the circular path surrounding certain large stupas. This prevented the superfluous layering of the tiles which the use of rectangular tiles would have caused. In Burma, we see numerous flat tiles on the Pagan site, for example, on mound No. 2306 of P. Pichard's inventory[80]. More rarely, we can see tiles featuring a side-by-side assembly (fig. 67) on site No. 2303. In Thailand, flat tiles have been widely used since at least the 8th century. Since the 18th century, the flat tile has often been enamelled and used to cover large structures. However, probably due to Khmer influence, the buildings have roofing with gutter tiles, and butt-end tiles with circular siding inspired by the Chinese style. Gutter tiles have been used for the longest period of time in Cambodia, since at least the 9th century on the Roluos site. On this roofing, the gutter tile and cover strip are differentiated (fig. 68). Contrary to the Chinese-style tiles, the siding is set above the gutter line of the structure. In India and Ceylon, builders also used the gutter tile, but the gutter and cover strip were not differentiated. These tiles were generally of mediocre quality. To better ensure a watertight quality, they were coated with curdled

[80] P. Pichard, op. cit. Vol. 8, No. 2306, pp. 238-239; No. 2303, p. 237. These two paragraphs are illustrated with drawings and photographs of the discovered tiles.

milk after being laid down as roofing. In Ceylon during the Dutch occupation[81] (17th-18th centuries), the Flemish purlin was introduced and met with great success. Exceptionally, from the 12th century and more generally later on, the roofing tiles were enamelled to improve their quality, as was the case in the royal palace of Angkor; enamelled roofing tile kilns were discovered on the Kulen plateau. The colour of the tiles was often modified by fires. Khmer tile makers rarely made specific tiles, with the exception of a large variety of tiles used in the décor of the architecture. Over the gutter line, the butt-end tiles close the opening of the valley tiles. These valley tiles ensured the space necessary for the roofing of two perpendicular elements (fig. 69). The bedding was often covered by very large tiles that completely covered the ridge and supported a ridge finial.

In Indonesia, the use of roofing tiles before the 13th century was rare. Those discovered in Sumatra near the Kambang Unglen site are of Chinese origin. There are no roofing tiles represented on the raised surfaces of Borobudur. However, they are represented on the buildings of the raised surfaces of the Prambanan temples (for example on the represented building No. 31 of the western balustrade of Candi Vishnu). These are wide flat roofing tiles with a cut-off lower edge. This form was used until the 14th century in Mojopahit. It was followed by small flat tiles with a catch. Finally, in Java as early as the 15th century, a Chinese tile from the Tang period[82] appeared. This double-curved tile (fig. 70) gradually replaced all other forms in Insulinde. Many forms were created to cover the hip rafters and ridges, especially in Mojopahit (fig. 71). For example, ridge finials were linked to the roofing supported by a radiating framework; in this case, the ridge finial protects the newel. This latter composition is very common in Bali (fig. 72). Bedding crowning pieces were made for roofing composed of flat tiles (fig. 73/A); this can be seen all throughout Indonesia.

Purely ornamental pieces were made in Burma, where blue ceramic tiles were inserted in the ornamental outline of the foundations of

---

[81] The Dutch probably adopted this type of tile after seeing the Tang tile, which looks similar, during their occupation of Java.

[82] Double-curved tiles are represented on the raised surfaces of the caves of northern China explored in 1907 by Ed Chavannes, *Mission archéologique en Chine septentrionale*. Paris 1913, fig. 315 and 340. This form was seldom used in China, although it is widely used in Japan.

certain monuments, for example in Mingalazedi. In Thailand, roofs are decorated with large pieces of terra cotta. The most impressive are the ending pieces that decorate the roofing peak, some of which are up to 1 metre tall (fig. 73/B). In Cambodia, the bedding is often decorated with a terra cotta purlin line (fig. 73/ C and D).

CHAPTER FIVE

STONE AND STONECUTTING

Stone has been used as a construction material since prehistoric times, particularly in northern India, in Baluchistan. The desert area of this province is covered in stones, usually broken apart due to harsh changes in temperature. As it was difficult to find mud cleared of its various elements, builders began to collect the largest stones they could find to construct buildings. One of the oldest structures uncovered is the silo[83] of Nindowari (Ph. 39). It is made up of rectangular caissons from the third millennium. The biggest stones are bonded together with mud and packed with smaller elements, forming what was called "mottled bonding". This bonding technique was used in construction for a very long period (until the 9th century A.D.) over a considerable area of northern India, including important sites such as Taxila. It is normal that over such a great period of time, over three thousand years of usage, the technique evolved. Sir John Marshall studied this changing process regarding the group of Buddhist monuments of Dharmarajika[84], from the 1st century A.D. to his day. At first, the stones were placed in any order. Then, little by little, the builders tried to establish more or less horizontal courses and attempted to use stones of the same size. Despite the fragility of the technique and the attentive care the set-up of the stones called for, builders still managed to construct corbels covering the paths in the That I Bai monastery (Ph. 40). Under the influence of the Bactrian master builders and despite the technique's lack of flexibility, the builders implemented the support of a very complex architectural décor of Corinthian pilasters and blind arcades at the base of the Chivaki stupa (Ph. 41). Under

---

[83] The site of Nindowari was excavated by Mr. J.M. Casal (1963-65). This silo is connected to dams (the "Gowar Band"), also constructed in stone using the "mottled technique", but it is likely that the current state of these latter structures are more recent than the state of the silo.

[84] Sir John Marshall, *Taxila, an illustrated account of Archaeological Excavations carried out at Taxila*. Cambridge University Press, 1951. The description of the site of Dharmarajika is the subject of chapters 10 and 11 of this work and the evolution of the technique is treated in Vol. I, pp. 248-49 and illustrated in Vol. III, pl. 55.

the same influence, the unburnt brick monument at Surkh Kotal was covered at least partially with a white limestone siding (fig. 74). The quarry from which it was extracted was not found, but it was most likely close to the site as the calcified outer layer was not damaged and was thus formed after the cleaning down[85]. The siding bond is very finely done. It is adjusted on the courses with a few uneven sections. At Taxila, under the same Bactrian influence, builders mixed courses of cut stone with courses with mottled bonding as early as the 3rd century. At the same time on the Sanchi site, builders began to construct in cut stone (stupa No. 3).

Stonecutting Techniques

Stereotomy is often derived from brick techniques. For example, stone corbels are sometimes sculpted with useless steps as the use of stone makes them unnecessary. Outside of India, it seems that builders did not construct with mottled bonding. Instead, they used stones collected from mountainous regions, which gave an even colour to the buildings, or from riverbeds, which furnished stones that differed depending on the areas the stream flowed over. In Java, the Candi Borobudur was constructed with stones collected from the bed of the Kali Progo, which in times of flooding, washed along great quantities of stones which were of different origins, but all volcanic: andesite. The stones underwent stone scabbling before being taken to the worksite. The preparatory cutting seems to have truly taken place at the stone collection site to make the transport easier. However, during the restoration work and the different searches conducted along the rivers of the site, a stone scabbling site was not found. In Laos, Vat Phu, the large temple near the city of Champassak, was constructed at the foot of a sandstone cliff from which great blocks were taken. These were cut first for the construction of compartments composed of sandstone slabs, and then for the construction of the monument. The use of the blocks broken off from the cliff involved

[85] This observation should be compared to an instruction in the *Mayamata* concerning the use of stones: "The sage rejects a stone licked by the wind, by blazing sun or by fire", which means that builders rejected stones with a calcified layer already on the surface and which would be altered by stonecutting. *Mayamata op. cit.* Ch. 33, § 33-4-5, A. This is all the more remarkable as at the time when the *Mayamata* was written, the stones used were granite or laterite, without a calcified layer.

a technique probably borrowed from India. On the surface of the block to be removed, builders drilled a line of furrows in which they inserted wooden chocks. They then thoroughly drenched the wood, which made it expand and thus break apart the stone according to the alignment of the chocks.

To construct the compartments, the slabs are implemented using a technique derived directly from carpentry and featuring tenons and grooves. This technical change is unique to Southeast Asia: in India, under the major influence of the Greek kingdoms of Bactria with their stonecutting, in Ai Khanum, and later in Taxila to shape column bases with sculpted profiles. This had a great influence on Kushan art, where for example the décor of unburnt brick structures of Surkh Kotal was made with cut stones imitating the Corinthian order. This led to a few inconsistencies; the base of the pilasters rest on the ornamental outline of the base of the wall instead of being set into them[86] (fig. 74). When the Pallavas began to cut sanctuaries into the mountain, they maintained pillars in them in an attempt to bring together both origins, wood and the Greek stone monuments. This resulted in true formal repetitions. In Taxila, on the Sirkap site, one can see the same type of pilaster as at Surkh Kotal for many platforms connected to stupas. In particular, the base of the pilaster rests on the base of the wall instead of being set into it. The courses are set at approximately the right height. For example, the height of the base of the pilasters corresponds to that of a course of the wall, even though the usual height was significantly superior[87]. The building on which the Greek stonecutting technique is most apparent is the temple of Jandial[88], particularly on the ionic base and capital of the porch (fig. 75). What is most striking is that the rest of the building is constructed with mottled bonding. One can see this combination at Sanchi, notably on the buildings of the 7th century that make up Buddhist monastery No. 36[89], where the walls are constructed with regular bonding for the base and mottled bonding for the rest.

At the same time, during the reign of the Pallava dynasty, builders

---

[86] The influence of Greek art would be highly significant in India, but led to many inconsistencies. For example, the pillars of the Pallava caves of Tamil Nadu copy a Greek column with a sort of architectural pleonasm: a double capital.

[87] Taxila op. cit. Plates 28 and 29.

[88] Taxila op. cit. Plate 44.

[89] Sanchi op. cit. Vol. III. Plates 123-124.

began to construct sanctuaries in cut stone, first in granite and then in sandstone. Granite was first used in conchoidal slabs extracted by fire for the Kuram and Kalampakan temples located in the north of Kanchipuram. The granite slabs were placed on edge in two parallel rows. The space between the rows was filled with bricks and notably brick debris. Above this first course, which formed the foundation, the superstructure was constructed using the same method but with a course of horizontal slabs set between two courses of vertical slabs[90]. The process of extracting the granite with fire, although seemingly little used, is still in practice today. We observed (in 1972) the process in the same region in Gingee used to create slabs to cover a small canal. From the end of the 7th century onwards, it seems that granite extraction by splintering was the general method. As we saw earlier, it was first used for irregular blocks in India, and particularly in Mahabalipuram, and then quarries were opened. The quarries that we visited are located in Tamil Nadu in Panangudi (pink granite) and Kudumalai (grey granite); both are in the district of Pudukkottai. The use of the granite involves a splintering technique with dampened woodened chocks. Before being sent to the worksite, the slabs underwent stone scabbling at the quarry into parallelepiped blocks that were small enough to be easily transportable. This was done with a tool that is similar to the European "scutch", a sort of hammer with a sharp point on one side and a stone hammer on the other. On the worksite, the work is continued using a punch and finished on the surface with a bush hammer. This tool is composed of a small cubic prism of about 6 cm on each side. One of its sides is covered with nine diamond points. This side is set on the stone and the bush hammer is hit on the other side with a stone hammer in which the handle extends from the head (contrary to the usual hammer in which the handle is perpendicular to the head). From the 9th century onwards, although granite was not abandoned altogether, the use of sandstone became more generalised, particularly in the region of Kancipuram. Sandstone is a stone with distinct beds. It was often extracted near the construction site. The block of stone was first outlined by a thin ditch cut out with an axhammer and then removed following the line of the bed. This method was also used in Southeast Asia, in particular amongst the Khmers. It is likely that the recommendations of the *Mayamata* that we mentioned above were used

---

[90] Photographs of the Kuram temple are reproduced in *Encyclopaedia of Indian Architecture, South India, Lower Dravidadesa*. Oxford University Press, 1983. Pl. 49 and 50.

for sandstone. Once removed from its bed, sandstone loses the water that it contains due to evaporation. It brings to the surface a part of the cement that binds together the grains that make up the stone. This cement is left on the surface of the block and forms a hard crust after a month or two. Therefore, stonecutters must cut and sculpt the stone as quickly as possible once it is removed. This is quite inconvenient as if the stonecutters wait too long to sculpt, not only will the hard crust be difficult to cut, but it will also be destroyed. This crust will not reform if the evaporation is completed, which is particularly obvious when the stone is reused. As the reshaping destroyed the protective crust, erosion is quick to follow and particularly destructive. In Kerala, also starting from the 9th century, builders extracted laterite using a process very similar to that used for the extraction of sandstone. However, the ditches encircling the blocks were made with the sole tool of the quarrymen. This resembled an axe with a handle measuring 1.20 metres in length. The other side of the blade featured a sturdy head (Ph. 42). After being traced out, the block was removed using this same tool. This left a stair-shaped cut on the face, used to limit the tracing to two sole ditches (Ph. 43). This form of quarry operation spread throughout all of southern Asia, and particularly in Cambodia.

The setting up of the granite blocks on the structure was usually done without a binder as the weight of the blocks ensured the structure's stability. This is particularly striking on the first Pallava temples constructed, for example in Panamalai (fig. 76). In this case, the space between two courses on top of each other was reduced to several centimetres, but the weight was sufficient to ensure the stability and support of the internal filling. This principle can also be seen on the Cola temples where the cut of the courses is sometimes very complex with diverse forms of convex and then concave curves (Ph. 44). In these cases, consistency is ensured by the weight of the stone blocks.

In Ceylon, as early as the 8th century, builders came up with a stereotomy directly inspired by carpentry. The temple of Aukuna was constructed with the method of two sidings enclosing a filling level. However, to maintain the spacing under the interior thrust, the builders set up devices uniting the two sidings with a complex assembly (fig. 77)[91]. The most remarkable effort made by the master builders

[91] This montage evokes the process described in the *Mayamata* under the name of Sarvathobadra. The orientation and implementation as they are indicated in the treatise can only rarely be used, see *Mayamata*, op. cit., p. 91 and fig. 9.

of Ceylon was the invention of forced chocks at the end of the 7th century. On the base of a support wall of one of the sanctuaries built on the Mihintale hill, we can see a chock forced in the horizontal plane (fig. 78). This technique, the most original of the entire region, became very widespread, first in Java from the 8th century onwards, and then from there to Cambodia in the 9th century. Finally, in the 10th century, probably spreading from Java or Sumatra, the technique reached southern India.

In Java, builders continued to collect stones to construct temples, even when they were very large. In the 13th century, for the principal sanctuary of Singosari, the builders did not use a quarry. However, it is likely that to prevent the uneven appearance that the walls of Borobudur have under their plaster, the builders attempted to collect stones of the same origin. For example, the Prambanan temples are almost the same shade of black, even though the stones were collected. The oldest monuments of the plateau of Dieng or Gedong Songo are built on a volcanic plateau on which the collection of stones was a simple task. As the buildings were small, the stones were set down flat without a binder in horizontal courses with regularly alternating joints. The roofing of the cella was done using a vault with a very regular corbel (Ph. 45). The door lintels were often protected by a discharging cell of the vault of which the intrados was sometimes recut to form a regular curve (fig. 79). The spread of Buddhism in the 8th century in the western part of the island brought about the construction of many monuments, including the Borobudur (fig. 5, X, second stage of construction). The techniques used were partly derived from Ceylon. For example, starting from the 45th course, the builders began to force several chocks whose number rose gradually as the worksite progressed. At first, the stereotomy[92] was very simple. The stones were laid down flat, connected with stone grouted hooks shaped with double dovetails (fig. 80). Some of these stones were cut at a right angle to the horizontal plane (fig. 81), but despite the size of the worksite, there was no systematization for the stonecutting. When the monument reached the 56th course, there was a major collapse which knocked down the last twenty courses put into place. When the work was started once more, it involved a new project that featured major reinforcement of

[92] I have described the construction of Borobudur in *Histoire architecturale du Borobudur*, EFEO, 1977, Ch. IV.

the base foundation (fig. 5, V). Technically, the builders used the same procedures, even though they had proved faulty. However, this time when they reached the 64th course, the master builder used a new cut for the siding stones over the entire perimeter of the structure. They were now cut at a right angle to the vertical plane (fig. 81,e), which increased the consistency as the horizontal courses were bound together. The paths between the galleries were covered with a corbel whose support was ensured by a crossette keystone set into the upper course of this corbel (fig. 82). This very clear breakthrough, used until the 14th century in Java, did not evolve. It might have led to a better use of the properties of stone, but it remained short-lived.

Towards 830 in Java, there was a new wave of Hinduisation that could be seen in a major construction campaign. It was also evident in the introduction of new techniques that were quite common in India, in particular that of the double siding surrounding a filling bonded together with mud or a mortar of any quality. This procedure became very widespread. It can be seen on the restorations of Borobudur (fig. 5 Y) and on the entire structure of Prambanan (Ph. 46). However, for the support walls and the elements constructed in hollow blocks, the assembly mode was improved with a long groove dug into the upper level of the course inset with a tenon on the lower level of the stones of the upper course in Prambanan. Despite their simplicity, these techniques allowed for large constructions and sometimes true construction feats[93]. One example is the restoration of Candi Kalasan at the end of the 8th century. This monument, originally built on a square plane, was completed with a large crown. This was done using four squinches as corbels (fig. 83) of which several courses were forced into place with chocks. In the previous chapter, we saw that Javanese bricks were often of poor quality, which did not allow for the construction of customary architectural elements such as stairs. This explains why stairs were covered with a sturdy andesite stone (at the Candi Gambar in East Java). Andesite was not the only stone used in Java; builders also extracted tuff. There are many quarries, particularly on the Ratubaka plateau near Prambanan. These are enormous worksites that were used for a long period of time. They feature working faces measuring over thirty metres high with access stairways cut into the

93 A taste for technical feats was completely common in Java. For example, at the Candi Lumbung (near Prambanan), on the side walls of the chapels, we can see large blocks whose siding probably called for great attention that was not necessary.

rock. Although this stone was of mediocre quality, it allowed for the construction of large structures such as the walls of enclosures, as was the case for the Candi Sewu and the Candi Sambisari.

It is likely that the Javanese Sanjaya dynasty spread its suzerainty to the Khmer land. In these conditions, it is not surprising to see Javanese techniques featured on Cambodian monuments. The most interesting borrowed technique was that of the forced chocks. From the middle of the 9th century onwards, the Khmer master builders made excellent use of this as can be seen, for example, in the construction of the pyramid of Phimeanakas in Angkor. When the links with Cambodia began to unravel[94], the new Javanese techniques were not taken up on the Khmer worksites. The double siding is completely lacking in Cambodian architecture. The forced chock technique also spread from Java or Sumatra to India during the Cola period in the 10th century. It was used on the great temples of Tanjavur and Gangaikondacholapuram, but from the 13th century onwards, it was used more and more rarely.

The Khmers seldom used collected stones. The stones of Vat Phu were used simply because the site was located at the foot of a cliff. Construction stones in Cambodia are extracted. There are many sandstone quarries on the Kulen plateau or in the surroundings of certain temples such as Beng Melea. Sandstone of many qualities was used, principally arenite[95] for the great temple of Angkor Wat, for example, but also harder sandstone such as greywacke for the temple of Ta Kev (greywacke quarries have still not been discovered). Apart from sandstone, laterite was also used and extracted. A large quarry was discovered near the temple of Bantay Srei of which the working face measured about 3 kilometres. There is no difference in the cutting technique between laterite and sandstone. However, taking into account the formation of a crust due to the evaporation of water of the quarry for the sandstone, there is a major difference in their use. While, for laterite, it is easy to make a stock of stones ready for use, the same cannot be said of sandstone. Sandstone must be used as rapidly as possible and its reuse is almost impossible for sidings. This fact was forgotten by the Khmers at the end of the 13th century. We

[94] The Khmer king Jayavarman II freed himself of the suzerainty of Java and from then on, Khmer construction evolved in an independent manner as of approximately the year 850.

[95] Arenite is mostly grinded in order to have excellent joints

can see numerous restorations with reused stones, which had disastrous results for buildings on which the protruding surfaces quickly eroded away. This can be seen in the last operations of the terrace of the Leper King or on the reliefs of the north-east corner of the gallery of the first floor of Angkor Wat. These reliefs were sculpted in the 16th century on a wall built in the 12th century. In other words, the sculptors had a harder time finishing their work and they destroyed the protective crust that could not be reformed as the quarry water had already evaporated. As a result, the work quickly eroded.

The corbels that cover the galleries of Angkor Wat feature a particular stereotomy. Each element of the roofing is cut in a double right angle whose planes are tilted towards the exterior (fig. 84). This ensured, on the one hand, better coherence to the structure and on the other hand, the runoff of part of the rainwater caught in the vault. This technique has its disadvantages as the slightest movements of the structure deform the corbels. The implementation of the vault was effected from two sides of the sections (the great length of the galleries made this first operation necessary: the division of the vault into sections) by reversing the direction of the stones at right angles, which at the point where they came together, were blocked in place by a keystone.

The use of wedges gradually dropped out of use, which can be seen in Cambodia in the Bayon (early 13th century) where the stereotomy is rather consistent. However, it seems that all the teams working on this prestigious worksite did not use the technique. For example, on the southern wall of the first gallery, the work was divided into different panels. On some, wedges were used judiciously, whereas on others, they weren't used at all (fig. 85, the wedges are marked by the letter X). In India, as well, during the 13th century, the use of wedges was not well understood. Thus, the steps of the stairways overhanging the entry pavilion and the second enclosure of the temple of Darasuram are blocked by wedges (a and b of fig. 86) that are layered over each other.

From the middle of the 16th century onwards, Mogul architecture became widespread in India. Before creating original monuments, the master builders copied the Timurid architecture, which they translated from brick into stone. In general, two types of stones were used: pink sandstone and white marble, whose extraction was perfectly handled. Most of Fatehpur Sikri, under construction from 1571 to 1585, a remarkably short period of time for such a large ensemble of varied shapes, was built using techniques that imitated those used for wood—without adding innovations. However, under the reign of Akbar's father, Humayun, a Timurid roofing form was introduced to the region: the double dome. One of the very first examples is in Delhi: the Sabz Burj (circa 1540, "the green tower", fig. 87). Master builders in all climates have always aimed to harmonise the inner and outer appearance of the structure. Usually, between the aspect of the intrados of a vault or that of an interior chimney of a temple and their outer appearance, there are two views which seem irreconcilable. The master builders attempted to correct this difference. The simplest and most widely used solution involved setting a ceiling in the interior space, thus removing the appearance of the chimney of the intrados. This form was used all throughout southern Asia. The trace of one of the oldest ceilings can be seen in Sambor Prei Kuk in Cambodia (Ph. 35) in the octagonal towers. Sandstone hooks inserted in the brick walls of the towers supported a ceiling. In Java at the Candi Kalasan (fig. 83) above the top of the squinches when the plane becomes octagonal, a cornice held up a ceiling which hid a chimney. Sometimes, builders tried to harmonise the intrados and the exterior appearance in Delhi. One of the first mosques constructed under the reign of Akbar, the Khayr al-Manazil mosque ("the most beautiful dwelling", built in 1562), has a prayer hall with a central section covered by a dome. To correct the displeasing aspect of the drum, it is divided into several parts above the square plane. With the implementation of four squinches, builders moved from a square plane to an octagonal one to a 16-sided polygon, finally obtaining the circle of the dome.

This latter solution was seldom used as it obliterates the outer aspect of the extrados of the dome, which is hidden by the superstructures of the building. Instead, builders used a technically complex form derived from the Sabz Burj that reconciled the two aspects of the monument. The biggest success of this composition is the famous Taj Mahal (finished in 1642, fig. 88). The external drum of the dome is high enough to clear it entirely of the superstructures constructed on the plateau covering the body of the building from which the internal dome barely emerges. Following the request of the emperor Shah Jahan, the construction of this monument was the subject of a detailed report drawn up by Mulla Abdul Hamid Lahori[96]. The report indicated that the

[96] The details of the construction and repairs of the Taj Mahal are taken mostly

foundations were lowered beneath water level, that the masonry of the foundation is made of limestone rough stones bonded with lime, and that the structures above-ground are in brick faced with white marble. The quarries where this marble was taken from are located in the region of Jaipur, in Makrana and Raiwala. The pink sandstone quarries are situated in Fatehpur Sikri and the immediate surroundings of Agra. The author points out that, given the proximity of the river, the foundations were built like port docks. The monument required repairs as soon as the work was finished, particularly on the dome and the vaults of the mausoleum. These repairs were done in 1652 and they continued through modern times.

This type of structure did not expand to Southeast Asia where the master builders of Muslim architecture continued to use existing shapes and techniques.

---

from the article by Madho Sarup Vats, "Repairs at Agra and Fatehpur Sikri: 1944-49". *Ancient India, Bulletin of the Archaeological Survey of India*, No. 6, January 1950, pp. 91-99.

CHAPTER SIX

BINDERS AND PLASTERWORKS

As soon as builders began to use unburnt bricks, they also made use of a binder to attach the bricks together in a consistent manner. This binder was composed of sifted soil and water often—but not always—mixed with a plant element. In northern India, this element was generally rice husks. Builders used this same mixture as a plasterwork, which gave consistency to the structure, whether unburnt bricks or pisé were used. In some cases, this plasterwork itself was covered. For example, for the half-columns that decorate the palace wall in Mundigak (fig. 54), builders added a layer of lime[97] to give them a white colour, or finely ground terra cotta to obtain the red that covered the door jambs. The complex décor of the enclosure of Surkh Kotal (fig. 55) was coated with soil mixed with straw chopped into very fine particles. In Lashkari Bazar, the plasterworks on mud were of various compositions. For example, the décor of the Moorish arch, present throughout the site, was coated with pisé or a lime plasterwork depending on the location of the site.

Lime

The invention of lime plasters was perhaps accidental; limestone was probably charred accidentally in a fire, and then builders tried to voluntarily reproduce this accident. Lime plasters are now prepared in the desert regions of northern India, using a very simple process with uncertain results. Even when the lime obtained is satisfactory, there is often much unburned stone waste that remains. First, the limestone is collected and broken up into pieces measuring about ten centimetres in length. Then, the baking space is prepared: a circular space 1.50 metres wide and 15 to 20 cm deep is dug in the ground. Then, the fuel is gathered. In Baluchistan, builders used camel excre-

[97] In this precise case of Mundigak, this was one of the first uses of lime during the third millennium.

ment that was dried and then mixed with a small amount of mud and chopped straw. The stones and fuel are mixed together and then placed in the prepared space, forming a cone of about 1.5 metres above the ground, topped by a layer of fuel, placed at the base under the mud. In this base, two or three shafts are left open with which the draught can be heightened using bellows. The baking lasts up to three days. The fuel is renewed from time to time, but the oven is left intact until after the cooling process. In southern India, builders made semi-permanent lime ovens (Ph. 47, the structure of the foreground is ready to be filled with limestone to burn and in the background, the oven is put out and in the process of cooling). In Bali, the lime plasters are prepared from pieces of the coral reef along the coast, collected by divers. The burning process is much quicker, sometimes only 24 hours. Also in Bali, shells were burned but only to prepare the small balls of betel. This ensured very rapid burning.

The use of lime was widespread in all of southern Asia. Special ovens were even made for its fabrication. Certain regions were preferred for their deposits of limestone capable of being burned, for example in Java, near the city of Bandung.

In Bactria, the Greeks used the plaster that they had been using for a long period of time. As they used it everywhere, they modelled ornamental outlines in stucco (plaster mixed generally with marble, although in southern Asia builders mostly settled for limestone crushed into fine particles). This deeply influenced the techniques of the sculptors and master builders in India. Between the Oxus and the Indus, all the important architectural remains, called Greco-Buddhist art—in the region are constructed in bricks coated in plasterworks. Analyses of the material used in Greco-Buddhist statues showed that calcium sulphate made up 82% of their composition, for only 0.75% iron, and that the remaining 17.25% were insoluble residue[98] (i.e. crushed limestone). In other words, these statues are mainly made of plaster. Although the plasterwork of the structures does not seem to have undergone any particular analysis, it is probably also plaster and not a lime coating.

As early as the 2nd century, the caves of Ajanta in India were decorated with large paintings of which many remains exist today. This décor is set over a stucco coating made with lime. The application of

[98] J. Hackin, *L'art Bouddhique de la Bactriane*, Kabul, 1937, p. 14.

this coating on the ceilings must have called for great finesse as in this case quick lime had to be used for the coating to set rapidly. Two parts lime are mixed with one part sand and then water is added. Once this is done, the workers have only about twenty minutes to spread the mixture, which means that the painting was not set into the coating. For the ceilings, a sort of tempera painting was applied to a layer that was already set and dry. For the inner and outer vertical walls, workers could use partly slaked lime which set at a slower rate. The temples of the Pallava period were coated and painted. Inside one of the cellas of the 8th century temple of Panamalai, the remains of a painting[99] were discovered. The condition of the painting reveals its application. The wall (composed of blocks of granite bonded together with a lime mortar) was first coated with a layer of lime and sand, then a second layer with sifted sand was carefully applied on top of that. It was over this latter preparation that the painting was created, before the coating set completely. This makes it possible to technically assimilate these paintings to the dry Italian frescoes ("fresco secco"). During the Cola period, there were genuine frescoes in the temple of Tanjavur, using a technique that seems to be unique to India in which mineral pigments were applied while the coating was still fresh[100].

The binders of the masonry are of extremely diverse quality. In general, the joints of Pallava masonry are slight, as the weight of the elements meant to be held together is such that the binder only serves as a wedge at the moment of application (fig. 76). When the Cola people began to build their great temples, as we have mentioned earlier, they used a stereotomy that allowed them to build with dry joints, without binders, but this was restricted to high-quality structures. More often, especially when the double siding technique was used, builders used binders of mediocre quality.

The walls of the Bengal brick temples are constructed with regular courses bonded together by a thin layer of mortar composed of lime and, instead of sand, brick crushed into powder. On exterior walls, the reliefs sculpted into the brick were generally left without coating (this was only added during repairs to hide the movements undergone by the structure). The extrados of the vaults is covered in a sturdy coating, one part lime for one part sand, applied with great care.

---

[99] This painting is reproduced in the book by Douglas Barrett and Basil Gray, *La peinture indienne*, Skira 1963, p. 39.

[100] *La peinture indienne*, op. cit., pp. 44-46.

Mogul architecture usually features no coating. The pink sandstone or white marble sidings are very carefully polished, making coating unnecessary and ensuring the beautiful appearance of the monuments. However, the siding had to be constructed with care and it is likely that its implementation called for a quick-setting lime mortar.

In the 16th century under Portuguese influence, exterior lime coatings increased greatly in quality, allowing for the creation of finely worked Baroque décor. There was a beginning period during which the leading Portuguese master builders were military engineers specialising in fortifications[101], which first lent a certain austerity to the structures in which the décor of the façades was limited to a single porch. After this period, as of the 17th century, ornate ornamentation made its place not only in religious buildings but also in palaces and on certain Hindu temples of the Goa region, for example at the temple of Shanta Durga in Quela where the external décor features blind arcades surrounded by Corinthian pilasters.

In Burma, the temples are built with a lime mortar of diverse quality. The walls were covered with a coating that was sometimes very thick (fig. 63). The detail of the ornamental outline was chiselled (rather than modelled) in this coating. After this step, the entire wall was covered with a lime whitewash that smoothed out all the rough surfaces left by the chiselling. Often, this whitewash was reapplied, even when the décor was not redone, but the detail lost some of its sharpness.

In Thailand, the Chaiya monuments are constructed in bricks set together with a binder of lime and sand, approximately made of one part lime to one part sand in rough amounts. The internal joints show much finer work than on the exterior, where the structural décor was prepared in brick before being covered with a plasterwork. This was probably intended to cover a large space and perhaps frame sculpted images. For Khmer monuments in Thailand, the binders and plasterworks of the décor use the same procedures used in Angkor, with no differences.

In Cambodia, in Sambor Prei Kuk, builders used a high-quality plant binder in very fine layers placed over the support beds, which were sanded down beforehand. There was probably no internal plasterwork. The walls were carefully polished, making the surface extremely even (Ph. 36). This is certainly the case in Angkor for the towers of

---

[101] Heler Carita, *Les Palais de Goa*, Editions Chandeigne, p. 15.

Prasat Kravan. The inner side of the walls of the central tower of this monument was sculpted with raised images (Ph. 48), which were then polished and painted directly over the sculpted brick. The range of colours is not easy to reconstitute, as only the red and black shades remain. Inside the temple of Prasat Neang Khmau, the walls are covered in paintings with the same inspiration as the raised images of Prasat Kravan. However, unlike the temple, there is no raised guide for the drawing, which was done directly on the polished wall[102].

On the exterior, the walls of the brick structures were hammered to ensure better adherence of the plasterwork. Before application, the décor was often roughly sketched out in a rather crude way. At Roluos, traces of paint were found inside the towers of the western row of Prasat Preah Koh. On the outside, the décor modelled in the stucco is very large with major raised sections (Ph. 4). It was put into place in several phases. First, builders set a mixture of sand and very liquid lime over the brick to stop up the pores of the walls. This made the brick less porous and slowed the setting of the second layer, which was thicker, as well as the third layer in which the décor needed to be modelled.

On stone, the laterite was sometimes coated. However, until the end of the 12th century, the sandstone was left revealed after being polished on site after construction and cleaning down. This ensured perfect consistency. However, in Angkor Wat, the raised décor of the first gallery was coated with a red-coloured preparation, which may have been a sort of lacquer. Then, a number of the décor elements were gilded, in particular the weapons and shields. Sometimes this preparation was even used for the backdrop. This mixed use may be explained by a change in the principle of gilding. At first, it was restricted to weapons, but later was used for the backdrop of the composition. Starting at the end of the 12th century, particularly in Bayon, all of the stone structures were plastered and the raised décor was painted. Several fragments of colour were discovered, revealing a rather rich palette. In addition to the black and red which had been used for a long period, fragments of blue and green were found.

In Java, due to the uneven nature of the stones (especially on the first monuments constructed in the 8th century, plasterwork was

---

102 Photographs of these paintings are reproduced in Boisselier, op. cit., Pl. LXIII.

necessary to give the masonry a consistent appearance. Borobudur was entirely plastered. This would also seem to be the case of Candi Sewu. The very tall statue of the central sanctuary, at least in its second state, was built with stone and thus plastered. The stone monuments of the 13th century from eastern Java were constructed with dry joints. The use of striking wedges, although not gone altogether, has become very rare[103]. Nonetheless, builders preserved from ancient stereotomy the use of the crossette keystone, which blocks the corbels of the Candi Singosari. These temples have an ornamental outline that is very carefully sculpted. Despite the rigour of these mouldings, it seems that there was no excess plaster applied. Instead, builders used a light layer of lime. A good example is the foundation of Candi Kidal, which directly supports the raised décor, with no reserve for a layer of plaster. The same can be said of Candi Singosari, where the Kala heads that crown the false doors of the upper level are sculpted in full detail—and builders never thought to completely model or sculpt one in the plasterwork itself.

The brick temples of the same period up to the end of the 14th century are built with a binder that is similar to that used by the Khmers. It is made of a vine sap mixed with palm sugar. Thanks to this very high-quality binder, builders were able to make large overhangs, in Candi Jabung for example (Ph. 38). This binder, combined with the indispensable sanding of the bricks, made it possible to not seal certain elements of the structures. For example, the inner side of the walls of the cella of Candi Jabung was not sealed. The cleaning down and then polishing of the wall surface alone produced a superb surface. (It is, in fact, this careful polishing that allows for the exclusion of plasterwork, as the plaster would not adhere to such a smooth surface). On the exterior, the plasterwork is set down following two main methods, mainly in several layers, generally three (it is difficult to evaluate the exact number of layers, as there were no doubt many repairs done). The first layer diminished the porosity of the walls and thus slowed the setting of the lime (like in Khmer regions). The second added volume to the ornamental outline. The third allowed for chiselling. When the ornamental outline was prepared in terra cotta, as is the case in Candi Gunung Gangsir (fig. 88), the housing of the pieces

[103] However, we can observe that on certain brick monuments, the Candi Bajang Ratu for example, the bricks were cut on the corners.

was set into the masonry. Then, the element was put into place with a plant-based binder, which served perfectly as the joints. Despite this, it is likely that the entire structure was covered with a layer of lime plaster to give it unity.

The Javanese brick monuments of the Muslim period were coated in three successive layers, often forming a very thick structure measuring approximately ten centimetres and often highly modelled. These three layers do not have the same composition. The first, set directly against the masonry, plays the same role as in the most ancient architecture: to eliminate the porosity of the structure. The second serves for the sketching out of the décor. Finally, the last layer of lime and fine sand brings out the details. The buildings that make up the entire Taman Sari in Yogyakarta are constructed in this way. Their plasterwork was sometimes modelled with the representation of a building covered in roofing tiles in full detail (fig. 89). For example, the bedding is precisely shown. The tiles set under the crown ridge are displayed. On the same site, a large plant décor takes up the spaces left by the architectural figures on the entire gable (fig. 90). This décor was painted, but unfortunately, all that remains is a few fragments of red, brown and black.

All throughout southern Asia, the climate causes irreversible destruction to the wooden structures. It is impossible to see the plasterworks that once covered the buildings, except for structures that were constructed recently. The assembly of wooden beams, apart from the techniques we discussed in Chapter II, are often simply beams tied together with rope. As of the 2nd century, the use of nails was seen in northern India and Bactria. However, they were not necessarily used for buildings as such. They were probably used for simple wooden structures. In Thailand, pagodas were often painted and gilded on the outside, creating a very rich décor (Ph. 10) while protecting the wood. This same technique can be seen in Cambodian pagodas. In both countries, gilding was often placed over lead using a procedure that added a shiny aspect to the décor ("gilt lead"). The treated sheets of lead were then attached to the wood. For less prestigious buildings, the wood was covered with an oily mixture and, more recently, with wax (Ph. 25 in Thailand, Ph. 29 and 30 in Java). On the interior of the building, the wood was left bare and generally polished; no other treatments were used.

CHAPTER SEVEN

OVERHAULS AND REPAIRS

As is normal with structures that undergo very long periods of use, overhauls and repairs are necessary. In an earlier chapter, we mentioned those of Taj Mahal, which began as soon as the monument was finished. The fashions had also changed and it became natural in the eyes of the users to revamp the structures, or at least their décor. For example, the temples of Gangaikondacholapuram and Tanjavur were completely transformed by this "makeover" by master builders of the Nayak period in the 17th century. P. Pichard writes: "The appearance of the entire tower is completely different: at Tanjavur, the mirror effect created by the number of kiosks is reinforced further by this lavish décor, which tends, on the contrary, to mask this desire for same-scale consistency and the progressive transformation of the shapes that characterise the tower of Gangai".[104]

Religious buildings sometimes change significance, which calls for readjustments that can be major. This is the case in Java when the Borobudur outline was changed. It was first Hindu then Buddhist. Another example is in Cambodia during the reign of Jayavarman VII where modifications of monuments probably followed the evolution of the kingdom.

Thus, there are two attitudes for architectural reworking: either, after damage or poor workmanship it is necessary to repair a structure and restore its former appearance, or on the contrary the master builder is looking to create a new appearance for the building. In the first case, we can estimate that although the work to be done was sometimes a great undertaking, this really involved the upkeep of the building. To do so, the builders used the same techniques as those of the original construction. In the second case, when the monument was meant to change significance, the new master builder had no reason to respect the former structures and sometimes showed great brutality in his works compared with the former structure.

---

[104] P. Pichard, op. cit. p. 89.

In about 795, when the Sailendra occupied the Dieng plateau, they undertook the transformation of a Shiva temple, the Candi Bhima, into a Buddhist sanctuary. To do so, they built a porch that not only protected the door but also partially hid the niches that sheltered the Dvarapala statues. These works were done without much consideration for the internal appearance of the former structure. For example, the light which probably lit up the linga through the support vault was blocked by the porch vault (fig. 79) with no concern for consistency; the niches are only partially hidden. However, on the exterior, the builders took greater care to link the ornamental outline. Despite the fact that the porch hid a side of the base, the master builder kept the octagonal[105] line, which is an exceptional feature for a Buddhist sanctuary. During the same era, the same Sailendra worked without much care when, remaining Buddhist, they changed doctrines from Hinayana Buddhism to Jina Buddhism. This called for major architectural changes for the insertion of the new statues. On the large sanctuary of Candi Sewu, these changes were most likely "financed" for the side chapels by individuals who commissioned the work from different master builders who did not all have the same technical skills. Thus, for the implementation of the doors, we can see ten or so different methods, some of which are damaging to the original sculpted lintels. In Borobudur, to set up the elements that were meant to hide the perspective effects, the builders dug the mortises in the second state of the structure in the middle of certain reliefs so that they could hold the tenons of the new stones without any concern for the damage (fig. 92). Sometimes, the cleaning down of the new structure caused damage that could have brought about serious consequences. Thus, due to the cleaning down of the corbel covering the access stairs of the second gallery, grouted hooks were cut longitudinally, which removed the effectiveness of the dovetail[106]. However, when the builders made a strong foundation that partially obliterated the former one, they had to take up the ornamental outline that was cut. This major work over 400 metres long was done with the greatest of care and the extension of the former torus with the curve that forms the just-built double-curved moulding is perfect. This also shows the

---

[105] We can see that the side of the square base of the foundation of the sanctuary is equal to the side of the circle inscribed by the octagon of the foundation, which was not modified by the addition of the porch.

[106] *Histoire architecturale du Borobudur*, op. cit. ph. 40.

builders worked with a curve which, on its own, could ensure that these difficult conditions maintained a continuous line. This third remodelling had another effect as well. By moving the point of departure of the base of the stairway, the builders widened the second gallery[107] and changed the architectural composition, which was most likely involuntary. Sometimes, restoration is very minimal and only serves to rework the fashion of the structure. For example, in the 13th century, probably when the builders were redoing the plaster of the monument, they worked over the reliefs and added, in particular, long tongues to the lions which seemed to support the represented temple. This was a change to reflect the times, a reference to India. In the 13th century during the Cola period, builders transformed the animals that seem to hold up the temple since the Pallava period. Instead of standing with their mouths shut, they now had open mouths with their tongues hanging out. In Borobudur, builders corrected relief E/1/10a so that it would be consistent with the model of the latest Cola temples (the temple of Amman in Darasuram, for example).

The Pallava temple of Panamalai was largely overhauled (fig. 93) with no concern for the architectural composition. As a first step, builders added a pavilion on columns in the front of the temple, which ran into the original temple but hid half of the inscription of the upper strip of the base. It is possible that this was not simple negligence and that the initiator of the new work was claiming the work as his own by hiding the inscriptions of his predecessor. The restoration took place after a period of abandonment due to the arrival of Islam in the region (a mosque is constructed at the foot of the hill that supported the temple, on the edge of the lake of reserved water to irrigate the neighbouring rice paddies). These new additions were in brick and carefully built, sometimes over former engravings (the enclosure, in its new version, is built over the former engravings in the rock). The restorations give indications of the changing techniques. Thus, the door of the west chapel was redone during the second work operation. At first, the opening was reduced from 1.25 metres to 0.80 metres. Then, builders most likely added a doorframe which brought the width of the door leaf to 0.65 metres.

[107] This widening was noticeably increased by the first restoration work, which the later work could not correct.

In Angkor at the end of the 12th century or the beginning of the 13th century, builders constructed a temple, the Bantay Kdei, linked to the great lake of Sras Srang. This was a true reworking of the entire space and the significance of the structures. First, the Sras Srang was constructed when the oriental baray began to show signs of sinking into the sand, to make up for the lack of water[108] to irrigate the space to the west of the lake. It was on this space that they decided to build the new temple. Thus, there was no more need to irrigate the land. The large lake was used instead and took on religious significance. To this end, the master builder drastically changed the dikes, which were coated in laterite mud and sandstone and a large landing stage (fig. 94) was built on the west side, approximately in the alignment of the new structure, which itself was remodelled at least four times. When the temple was set up for the first time, the superstructures of the corner towers of the first enclosure were probably limited to a cloister arc as a corbel. When this décor was no longer in style (it was maintained at Preah Khan in Angkor), the builders decided to crown the principal towers with a more elaborate form, in particular the corner towers of the first enclosure (towers 3, 5, 7 and 9 of fig. 95). To do so, the tops were levelled down at the level of the extrados of the corbel crossing (only keeping a single course of thickness). Over this crossing, they built a tower with false floors. On the central tower, the technique was different. The builders kept the décor, which was already a tower with false floors, and stuck the new décor over the old, often in a clumsy manner.

Overhauling was sometimes due to an error in the work process. For example in Candi Jago, in the east of Java, built in the 13th century, there was a layout mistake for the second register of the foundation of the principal sanctuary (ph. 50). Starting from the third course, the builders found the correct layout and rectified the plane by creating an

[108] The digs of B.P. Groslier between the Sras Srang and Bantay Kdei revealed the canal of water distribution which was constructed in laterite mud halfway up the barrier that made up the original edge of the lake. P. Courbin, *La fouille du Sras-Srang*, EFEO 1988. M. Masako Marui points out (in an article published in No. 10 of the review *Aséanie* dated December 2002, *La découverte de statues bouddhiques dans le temple de Bantay Kdei*) the discovery in the drilling D.U.08, located on the wall of Bantay Kdei, east of the third surrounding wall, of a small canal that was very similar, measuring 60 cm wide and 50 cm deep. This suggests that the installation discovered by B.P. Groslier, abandoned for the construction of the temple, was much more important. The canal discovered by the Japanese mission must be an installation that ensures a new distribution of water (see fig. 94).

overhang. We can imagine that the space left free over the two lower courses was filled in. If this was so, this was done with no fitting and the stones have now disappeared, which reveals the overhang—showing the layout error.

The overhauling inside of the sculpted caves was done to suggest an exterior shape. For example, in Karla, the sanctuary was cut into the rock sometime in the 2nd century. It consists of a room with a circular roof which carefully takes up the form of a wooden structure. To make the illusion even more believable, the builders inserted real wooden screeds between the cornice and the circular arch cut into the rock (C on fig. 96). Although it supports nothing, this superb framework must have been redone several times since its construction in the 2nd century. Its placement in the rock ensured the future of the sanctuary, but the concern for the image imposed many remodelling operations so that the illusion too would last.

Overhauling can also be part of a change in an architectural plan as an expression of a project. We believe this to be the case of Bayon: the anticipation of the conquests of the king expressed by the architecture of the monument, the evolution of religion and conversions to Buddhism described not only by the epigraphy but also by the architectural changes that had no technical reasons behind them.

Chapels were built in the courtyard (fig. 97) linking the external gallery to the second gallery which itself had been radically changed from a cruciform plane to a square plane. These chapels are built in a correct way with what seems to be embedding behind the wall with reliefs of the exterior gallery, which ensured the consistency of the entire structure. However, these connections with the internal gallery were only of mediocre quality and more importantly, no effort was made to hide the foundation of the internal gallery at the locations of the connections. The primitive structure could have reappeared at the moment of the destruction of the chapels.

This involved major work. It was almost like the construction of a new building rather than a reworking that had no technical reasons behind it. The monument certainly needed repair but in no way called for this amount of work.

Revamping is not limited to monuments; it can also apply to architectural texts[109]. For the construction of the Gangaikondacholapuram,

[109] I am much indebted to Bruno Dagens for his information, which this paragraph is based on.

when the master builder realised that the traditional layout methods of the octagon proposed in the treatises were not sufficient (see fig. 1 and 8), a new method was developed and used with success. When certain architectural texts were revised, such as the Mayamata, the new author incorporated a basic yet sufficient explanation of the new technique (the value of the side of the octagon compared with the side of the circumscribed square) in the former text, without actually modifying the text. In other texts, such as the Manasara, this information is not mentioned in such a fashion as to calculate the side of the octagon. In a general manner, this text does not seem to have evolved and the indications must be followed as they are. This is not always the case. For example, in the Silpa Prakasa, although this does not entail reworking or repairs, there are often indications that are sufficiently imprecise so that builders can take up the work without affecting the entire structure. Take for example verse 121: "By beautiful incision, they are made with two circles, one circle, or three circles. Having first determined the space, it should be divided into four parts." We understand that one single circle is necessary, but for the correct layout of the décor, it may be useful to opt for three circles, which would prevent overhauling.

It seems that there are no specific techniques for repairing monuments. Despite the abundance of repairs, it seems that builders were not eager to take up the work, even when it seemed urgent. Apart from the all-powerful ruler, it was surely difficult to find the financing for a monument, even when the monument offered a sense of prestige to the commissioner. For repairs, the prestige being less, the financial difficulties were even greater. Builders sometimes found solutions in strange ways. At the end of the 8th century, the Buddhist rites changed and the plans of the sanctuaries had to be cruciform to be able to shelter the statues of the five Jina, and thus certain temples underwent major work (the Candi Kalasan, for example). Yet, builders often settled for modest operations. At the Candi Lumbung, they simply added a door with one leaf which closed the central niche of the façades. At the Candi Bubrah, whose original state featured niches to shelter standing statues, the master builder had to use the space for the new iconography composed of statues of sitting Buddhas. He thus placed a sculpted element of a lotus taking up half of the height of each niche on which the sitting statue was placed. At the beginning of the 9th century, the layout rites were changed. It was no longer possible for the centre of the space where the temple was constructed to be covered by a structure. The overhauling of the temples would have called for considerable work and builders often renounced being consistent or simply abandoned the structures. There were a few exceptions, however. For example, at the Candi Badut (near Malang), the temple was left alone, but the builders redid the enclosure in such a way to empty the centre of the space.

In Java, as of the 13th century, the political conditions changed the attitude of the master builders in terms of repairs and overhauling. Despite appearances, the charter of Sarwadharma (1269)[110] marks a certain break between the power and the clergy for the upkeep of religious foundations. This radically diminished the number of new constructions and certainly increased the amount of less expensive repair work. The form of the work of the last overhaul of Candi Badut in the 13th century, which consisted of adding small structures to recreate a form consistent with the model of the times, was no doubt affected by this new policy. This is probably also the case of the last repairs of the reliefs of Borobudur, which date from the same period.

[110] The charter of Sarwadharma as studied and translated by Theodore G. Th. Pigeaud, *Java in the 14th Century*, The Hague, 1960, T III, pp. 143-150, translation ; T IV, pp. 381-390, comments.

CONCLUSION

This study has shown that the techniques in southern Asia are highly consistent. This is not only due to the strong influence of India on the region, but also to the determinism of the use of certain materials or their rarefaction. For example, the construction of roofs placed on screeds was abandoned around the 10th century, probably because the large vats necessary for their construction were no longer available. Despite the great variety of solutions, the placement of shearing points on a horizontal beam resulted in a resemblance to bent frameworks, no matter where they were used. This was also the case for "taut ridge" frames, no matter where they were used (fig. 98). The image is the same. It is the determinism of the materials and the technique that makes this so. In order for the ridge to be taut, the gables must be turned towards the outside and the wood used must have good resistance to traction. As with the screeds, when this technique was abandoned for large buildings, it was for the same reason: the disappearance of large vats. Builders attempted to keep the image of the taut ridge in buildings by introducing a bent framework inside the structure. In Cambodia, the builders even transposed the hoop roofing by sculpting the extrados of an overhanging vault in sandstone. These examples show us the extent to which construction methods[111] make their mark on the architectural shape of monuments and become more than a simple means. Thus, at the end of the 18th century, when the use of the radiating framework was replaced by that of the truss, the division into three elements, initially indispensable to ensure that the rafters weren't too long, was maintained. The technical consistency of the region is also marked by stonecutting. In its diverse forms, it was used from Ceylon to Java, and from there it moved to the Khmer region and finally headed towards India, with the improvements that the Java master builders lent it (fig. 99). The use of the forced chocks as well as the careful grinding of the joints gave the walls of Angkor Wat a perfectly even quality. It also gave sculptors the illusion of a

[111] When concrete was introduced to southern Asia, the flexibility of the technique removed many constraints and gave the illusion that the architecture was independent from the technique.

consistent rock surface, allowing them to sculpt bas-reliefs. However, when the sculptures were made in sandstone with a hard-set calcium deposit, the builders overlooked the constraint of working rapidly before the water of the quarry was completely evaporated.[112]

As it is clear what is possible and impossible to build, the mastery of the techniques allowed for the anticipation of the finished work. As a result, a certain distance was created in relation to architectural models. This was probably modest at first, but starting at the end of the 12th century, it grew much more prominent. This attitude towards architectural projects was particularly sensitive for the execution of perspective effects. In Angkor Wat, for example, where the current mastery of the technique is remarkable, it allowed builders to take into consideration the vision one would have of the monument after passing through the enclosures. The shapes appear independent from the technique, or rather its mastery makes it possible to seem independent.

This evolution of the mastery of an architectural work had consequences on the interpretations that one made of the tales and representations of creation, for example on the omnipresent ones of the churning of the milk sea. When the gods and demons undertake this operation, they do not think about all of the consequences of their actions. They did not truly plan for the results of the operation; the technique was not yet mastered and they were unaware of the birth of millions of Apsara. However, when the gods and demons are not able to move the Mandara mountain, Vishnu transforms into a tortoise and slips under it. On some reliefs, we can see a point on the top of the tortoise shell that facilitates movement. In this case, Vishnu at least anticipated a part of what his action would produce. In this way, the master builder is comparable to the god, because he too knows what his techniques will produce[113]. This is a new attitude in which chance plays only a slight role.

In the definition of architecture, when builders introduced the idea of an evolution of structures, it was no longer the master builder who anticipated the architecture in terms of the vision one would have of

[112] This is particularly obvious in Bakong at Roluos where all of the sculpted reliefs are eroded except on one stone, which was probably renovated.

[113] This is comparable to the attitude of Suger, the master builder of the Saint-Denis basilica, anticipating the plays of light that the imagined detail of the arches would produce on his work.

the monument, but rather the commissioner of the architecture who envisaged an evolution of the monument independent of the means, according to its significance. For example, when King Jayavarman VII had Bayon constructed, he most likely wanted it to be in the image of the ultimate divine work: the creation which was not accomplished in a single movement, justifying a great amount of reworking. In Java, it is the separation between the clergy and the royalty that changed the relationship of buildings and their significance. When the king no longer participated in the prestige of the construction of a temple and its upkeep was left to the clergy alone, builders no longer constructed works as major as Candi Sewu or Prambanan. In fact, technical changes were almost inexistent in Java after the 10th century. In India, the master builders wanted their works to be consistent with the directions of the treatises in which the fabulous dimensions were sometimes a far cry from reality. For example, in Hampi, the volume of the ornamental outline is designed with no thought for its layout; some stones are left abandoned halfway up the stairway, which was used as scaffolding due to the fact that the master builder had completely neglected the constraints of working with stone.

Nevertheless, although the case of Hampi is not unique, the best understanding of the limits of the techniques—in particular of stereotomy—led to buildings that often yielded to these constraints. For example, at the Preah Khan of Kompong Svay, in order to avoid complicated stereotomy, the builders used very large elements. These elements were not only used for the force that they represent but also to ensure faultless coherence of the structure. It is likely that after a great number of accidents, the builders realised that the understanding gained did not allow them to go beyond a certain height, or that the height depended on the width of the base of the building. For wooden structures, the proportions that the carpentry techniques called for would have been such that the laying out of the structure—although possible—would have been difficult to attain. Thus, the builders contented themselves with transposing the desired image in stone. The Cola designers of the great temples of Tanjavur and Gangaikondacholapuram generally made their monuments like wooden buildings. Although stone allowed for greater verticality, they did not want to stray from the model. Thus, the façades show a great number of pillars that were put in place despite the requirements of the stereotomy simply to give an image of a structure different from the actual one in place.

The architectural techniques of southern Asia not only allowed for the construction of monuments, but they also set limits. Builders quickly realised that it was not possible to build simply anything. Master builders invented artifices to make the user believe that what he saw was what he really wanted to see. In this way, the history of techniques and its possibilities lies at the very base of the history of architecture. The history of techniques is the history of difficulties overcome, not only in reality but also figuratively. The temple is materially the haven of the gods and has this image. As the stereotomy did not make it possible to construct a floating palace, the builders gave it an image that they deemed to be realistic. The palace does not float, but the garuda seem to hold up the structure, in Angkor Wat as in Penataran.

BIBLIOGRAPHY

Acharya, P.K. *Mansara,* second édition New Delhi,1979.

Casal, J.M., *Fouilles de Mundigak,* Paris 1961.
Casal, Jean-Marie, *La civilisation de l'Indus et ses énigmes.* Paris 1969.
Chavannes, Ed. *Mission archéologique en Chine septentrionale,* Paris, 1913.
Clément, Sophie et Pierre Clément, *L'habitation Lao,* pris 1990.
Coedès, george, *Inscriptions du Cambodge,* 8 volumes, EFEO, 1942–1966.
Coomaraswamy, Ananda, K. *Medieval Sinhalese Art,* New York, second edition 1956
Cunningham Alexander, *The stupa of Bharut,* New Delhi, reprint 1962.

Dagens, Bruno, *Mayamata. Traité sanscrit d'architecture,* Institut français d'Indologie, Pondichéry. 1970-1976.
Dagens B, Marc Le Berre et Daniel Schlumberger, *Monuments préislamiques d'Afghanistan,* Paris, 1964.
Dagens, B et M.L. Brazer-Billoret, *Le Rauravagma, un traité de rituel et de doctrine Sivaite,* Institut français d'Indologie de Pondichery, 2000.
Dumarçay, Jacques, *Histoire architecturale du Borobudur,* EFEO, Paris 1977.

Godard A et Y, et J. Hackin, *Les antiquités bouddhiques de Bamyan,* Paris, 1923.
Goloubew V, *Ajanta les peintures de la première grotte,* Paris 1927.
Goodman, W.L. *A History of woodworking tools,* London, 1964.
Gray, Basil and Douglas Barrett, *La peinture indienne,* Paris 1963.

Hackin, J. *L'art Boudhique dela Bactriane,* Kabul, 1937.
l'Hernault, Françoise, *Darasuram,* EFEO, 1987.

Leroi-Gouran, André, *Evolution et technique, Vol I L'homme et la matière, Vol II Milieu et technique.* Albin Michel, Paris 1943-1945.
Luce, Gordon H. *Old Burma – Early Pagan,* New-York, 1969 – 1970.

Mackay, E.J.H., *Further Excavation at Mohenjo-Daro,* New Delhi reprint 1976.
Marshall, sir John, *Mohenjo-Daro and the Indus Civilization,* London 1931.
Marshall, sir John, *Taxila, an illustrated account of Archeological Excavations carried out at Taxila,* Cambridge 1951.
Marshall, sir John and Alfred Foucher, *The monuments of Sanchi,* New Delhi, reprint 1982.
Mitra, Debal, *Buddhist Monuments,* Calcutta 1971.

Parmentier, H. *Art Khmer Classique,* Paris 1930.
Parmentier, H. *L'art du Laos,* reprint EFEO, 1988.
Pichard, Pierre, *Vingt ans après Tanjavur, Gangaikondacholapuram.* EFEO, Paris 1994.
Pichard, P. *Inventory of monuments at Pagan,* EFEO-Unesco, 1992.

Wheeler, sir Mortimer, *The Indus civilization,* Cambridge, 1960.

LIST OF ILLUSTRATIONS

Photographs[114]

---

[114] All photographs were taken by the author, unless otherwise indicated. They were taken between 1954 and 1987.

[115] Pp. 144-145 of P. Pichard vol. I, op. cit.

modern truss built over a radiating framework.

Pl. 20. Siem Reap, Cambodia, triangulation projects in a pagoda with three naves.

Pl. 21. Lombok, grain silo covered with roofing resting on screeds.

Pl. 22. Jakarta, Indonesia, framework of the canopy of a Chinese house.

Pl. 23. Panamalai, Tamil Nadu, north-east corner of the temple.

Pl. 24. Chiang Mai, barred windows of a pagoda with three naves and three breaks.

Pl. 25. Mandalay, Burma, longitudinal façade of a building on partially buried piles.

Pl. 26. Mandalay, remains of a pagoda sanctuary.

Pl. 27. Siem Reap, window of a guest house of a pagoda located within the enclosure of Angkor Wat.

Pl. 28. Kudus, Java, window of a 19th century house.

Pl. 29. Salé, Burma, relief of the crown of the cornice of the monastery sanctuary.

Pl. 30. Kudus, street in the old part of the town.

Pl. 31. Kudus, detail of a sliding door.

Pl. 32. Mudianpakhan, Tamil Nadu, west of Madras, brick oven.

Pl. 33. Pagan, Mingala zedi, stupa of the corner of the upper terrace.

Pl. 34. Pagan, barrel vault.

Pl. 35. Sambor Prei Kuk, inner wall of an octagonal tower made of brick, with sandstone hooks supporting the ceiling beams.

Pl. 36. Angkor, entry pavilion of the first terrace of the temple of Ta Kev, corbel covering the northern wing.

Pl. 37. Candi Jabung (eastern Java), eastern façade.

Pl. 38. Principal sanctuary of Padang Lawas (central Sumatra).

Pl. 39. Nindowari, Pakistan. (Photo by J.M. Casal), silo XLII of the silo.

Pl. 40. That i Bahi, Pakistan, inner door of monastery.

Pl. 41. Chivaki, Afghanistan, base of the principal stupa (photo from 1957).

Pl. 42. Kerala, quarry tool.

Pl. 43. Kerala, laterite carrier.

Pl. 44. Gangaikondacholapuram, stereotomy of the enclosure.

Pl. 45. Dieng Candi Bhima, corbel covering the cella.

Pl. 46. Prambanan, temple of Siva, southern wall of the first enclosure.

Figures

Fig. 37. Borobudur, represented architecture; the door is shown without a leaf.

Fig. 38. Borobudur, represented architecture A/ relief O/1/54, b. B/ relief N/1/86, b.

Fig. 39. Minangkabau house, axonometry, Sumatra, Indonesia.

Fig. 40. Bamboo bridge over the river Kanta, Siem Reap region.

Fig. 41. Sandstone window of the fourth floor of Angkor Wat, outer side.

Fig. 42. Sandstone window of room N of Preah Vihear, inner side.

Fig. 43. False window of Bantay Kdei.

Fig. 44. Wall of a Malaysian house of Seramban.

Fig. 45. Angkor Wat, false door with leaf, sculpted on the pavilion of the western entry of the exterior enclosure.

Fig. 46. Implementation of a wooden door leaf in a sandstone sanctuary dating from the 9th century. a/ pivot stone, b/ casement of the socket of the door pivot, c/ casement of the lock, the door being open, d/ support of the leaf, the door being open, e/ closing stile. Arrow 1 indicates the movement of the leaf after the presentation of the axis in the casement of the socket of the door pivot. Arrow 2 indicates the setting of the axis in the pivot stone during the putting into place of the socket of the door pivot. Arrow 3 indicates the setting of the axis in the socket of the door pivot.

Fig. 47. Pagan, temple of Ananda, western interior door, plan and elevation.

Fig. 48. Pagan, fragment of a fresco of the temple of Ananda Okkaung.

Fig. 49. Ceylon, one-leaf door of a pagoda in the surrounding area of Kandy.

Fig. 50. Ceylon, interior view of a door of the palace of Kandy and its setting in the masonry. A complete door, similar but with a wooden door frame, is on display at the museum of Kandy.

Fig. 51. Luang Prabang, door of pagoda.

Fig. 52. Bali, door of a house in the village of Ubud (1979).

Fig. 53. Angkor, temple of Ta Prohm, jamb of the northern door showing the casement of the repaired leaf.

Fig. 54. Mundigak, Afghanistan, plan and elevation of the wall of the palace.

Fig. 55. Sukh Kotal, Afghanistan, elevation of the décor of the interior enclosure wall.

Fig. 56. Barrel vault constructed with unburnt bricks.

Fig. 57. Arches made of unburnt bricks in Lashkari Bazar.

Fig. 58. Creation of bricks in Sawanti (village near Magelang, Java, 1977-1978) at the period of the drop in the rice paddy water levels. Legend: a, rice paddies at normal level before work; b, area where the paste is prepared; c, brick drying area; d, oven; e, rice paddies at normal level after the work.

Fig. 59. Perspective of the intrados of the brick vault of a temple of Bengal, "bangla" style.

Fig. 60. Darasuram, temple of Amman, cross section of the cella ceiling.

Fig. 61. Darasuram, axonometry of the intrados of the vault of the Amman temple cella.

Fig. 62. Western elevation of the Mingala zedi stupa in Pagan (based on the photogrammetric information of the IGN).

Fig. 63. Mingala zedi, masonry detail of the capping of the first floor, western façade.

Fig. 78. Stereotomy of the base of the foundation of a building in Mihintale (Ceylon). A, striking wedge in the horizontal plane; B, wedge put into place after the course was adjusted.

Fig. 79. Discharge arch above the lintel of a door at Candi Bhima in Dieng, Java.

Fig. 80. Borobudur, courses 41 and 45 of the internal base of the first gallery, implementation of double dovetail spikes, C, of the representation (based on pl. LII of *L'Histoire architecturale du Borobudur*, op. cit.).

Fig. 81. Borobudur, stereotomy of the internal base on the level of the third gallery of the 69th to the 74th course, with striking wedge, and angle bar in the vertical plane (based on pl. LIII of *L'Histoire architecturale du Borobudur*, op. cit.).

Fig. 82. Borobudur, passage from the first to the second gallery, crossette keystones.

Fig. 83. Candi Kalassan, Java, passage from the square plane to the octagonal plane of the cella (corbel trumpet arches).

Fig. 84. Angkor Wat, stereotomy of the corbel covering the gallery of the first floor and implementation of a keystone.

Fig. 85. Bayon, eastern wall of the first gallery, south side, stereotomy detail. The stones marked "X" are striking wedges.

Fig. 86. Darasuram, stereotomy of the northern staircase of the entry pavilion II/east, stones a and b are wedged.

Fig. 87. Delhi, Sabz burg, cross section.

Fig. 88. Agra, Taj Mahal, cross section.

Fig. 89. Candi Gunung Gangsir, Java, detail of décor set into the brick bond of the base of the sanctuary wall; the different elements modelled into the terra cotta are hatched.

Fig. 90. Yogyakarta, Taman Sari, elevation of pavilion 14.

GLOSSARY

Barrel vault–*noun* Demi-cylindrical vault section (fig. 56, ph. 34).

Bending, bending frame –*adv.*, *noun* Framework whose every element is handled with compression (fig. 19, 27, ph. 11).

Bond–*noun* A bonded wall is composed of elements specially cut to occupy a determined space. By extension, we use the expression "varied bonding" to signify elements that are not cut but rather chosen according to the space that they will take up in the wall (ph. 39, 40 and 41).

Closing stile–*noun* Cover strip attached to one of the door or window leaves (fig. 45 and 46).

Corbel–*noun* Structure whose elements feature a protruding surface (fig. 66, ph. 36 and 45).

Core–*noun* Central piece of a radiating framework (fig. 72, ph. 13).

Crossette–*noun* Crossette keystone: joint whose two sides feature a protruding piece (fig. 82).

Crown post–*noun* Vertical piece linking the apex of two principal rafters to the tie-beam (ph. 14).

Discharging arch–*noun* Arch placed in a wall to help support the underlying piece, generally the lintel (fig. 66).

Drill–*noun* Archimedes drill, carpenter's tool to drill circular mortises (ph. 5).

Fillet–*noun* Mortar plaster used to fill in the empty spaces of the roofing.

Gable–*noun* Capping of a wall whose apex carries the end of the ridge cap (fig. 32). False capping in south-east Asia: structure featuring a gable over the sloping side of a roof (fig. 25).

Hip rafter–*noun* Piece of wood that forms the corner of a roof.

Hip roof–*noun* Oblong roofing element, rounded on the small sides (ph. 17 and 18).

Intrados–*noun* Concave part of an arch or vault (fig. 59, 61).

Keystone–*noun*	Stone cut to ensure the blocking of a course (fig. 84).
Line of Jupiter–*noun*	Assembly of beams in a horizontal plane (fig. 13).
Panelling–*noun*	Wood surfacing of a slat (fig. 22).
Pivot stone–*noun*	Casement of the axis of a door leaf inset in the lintel (fig. 46).
Principal rafter–*noun*	Piece of wood that supports the purlins set into the base of the tie-beam and linked, at the top, to the opposite principal rafter (ph. 19).
Purlin–*noun*	Horizontal beam perpendicular to the trusses (ph. 10).
Rafter –*noun*	Each of the pieces that support the slats on the pitch of the roof.
Ridge finial–*noun*	Ornament decorating the crest or apex of a roof (fig. 72, 73).
Ridgecap, ridge–*noun, adj.*	Belonging to the ridge. A ridge slat or ridge piece. In this thesis, we have used the term ridgecap to mean "the beam forming the apex of the frame".
Riolet* wheel–*noun	Small wheel that serves to roll up the alignment cord of the carpentry pieces (ph. 5).
Roof angle–*noun*	Junction at the top of two pitches of a pediment (fig. 73).
Roof break–*noun*	Angle formed by the two sloping sides of a ridge roof.
Screed–*noun*	Curved roof truss (fig. 96, ph. 21).
Siding–*noun*	Exterior or interior side of a wall (fig. 74, ph. 35). The technique called "double-faced siding" is a construction method that uses two walls surrounding filling (fig. 77, ph. 46).
Socket of door pivot–*noun*	Piece of stone or metal receiving the lower point of the axis of a door leaf (fig. 46).
Stereotomy–*noun*	Shape and cut of stones (fig. 76 and 78, etc. Ph. 44).
Striking wedge–*noun*	Stone cut in the form of a prism and forced into a casement cut in a course (fig. 78).
String–*noun*	Newel of a staircase (fig. 64).

Stucco–*noun* Mixture of plaster and marble powder. We used this word in terms of its usage in southern Asia to signify a very finely sifted mixture of lime and sand that does not include marble or plaster, except sometimes in northern India (ph. 3).

Tie-beam–*noun* Horizontal piece joining two principal rafters or forming the base of a bending frame.

Trussed–*adj.* The truss consists of at least three essential pieces linked together: a tie-beam and two principal rafters; this assembly is, in principle, not bendable (ph. 19 and 20).

Valley–*noun* Inset angle formed by the intersection of two roofs (fig. 69).

Wall plate–*noun* Plate set at the top of walls.

Warped–*adj.* A warped wall or gable is a structure that is not in a vertical plane (wall: fig. 26, gable: fig. 35).

INDEX OF MONUMENTS AND SITES[116]

[116] The geographical location of the sites is shown on the map of fig. 100.

PHOTOGRAPHS* & FIGURES

* All photographs were taken by the author, unless otherwise indicated. They were taken between 1954 and 1987.

Ph. 1. Temple of Halebid at Karnataka, relief of the foundation of the principal sanctuary.

Ph. 2. Temple of Bayon, exterior gallery, west side, layout mark of reliefs.

Ph. 3. Temple of Preah Koh southwest sanctuary, south side, false door showing successive layers of plasterwork.

Ph. 4. Temple of Bayon, south exterior gallery, ease panel near the south central door.

Ph. 5. Borobudur, first gallery, relief 1/B52.

Ph. 6. Indonesia, Java and Sumatra. Archimedes drills (standing tools) and *riolet* wheels on the bottom.

Ph. 7. Preah Vihear, Cambodia, east entry pavilion, wall plate casement behind the cornice.

Ph. 8. Koh Ker, Cambodia, northern building of east entry, purlin casements of porch roofing.

Ph. 9. Chiang Mai, eastern façade of a pagoda with five naves.

Ph. 10. Swe-dan-gu pagan temple (No. 73 of the inventory)*, eastern façade, purlin casements of porch roofing.

* Pp. 144-145 of P. Pichard vol. I, op. cit.

Ph. 11. Village near Darasuram (Tamil Nadu), bent framework. (Photo of the Institut français d'Indologie de Pondichéry).

Ph. 12. Mahabalipuram, façade of the Draupadi ratha.

Ph. 13. Bali, Gjanjar, radiating framework.

Ph. 14. Bali, Klungkung, crown post supporting the newel of a radiating framework.

Ph. 15. Solo, Java. Central part of the framework of a *pendopo*.

Ph. 16. Mahabalipuram, façade of the Bhima ratha.

Ph. 17. Mahabalipuram, side façade of Nakula Sahadeva ratha.

Ph. 18. Kerala, framework of a building in half dome.

Ph. 19. Malaysia, near Lembah Bujang, mosque of Pangkalan Kakap, modern truss built over a radiating framework.

Ph. 20. Siem Reap, Cambodia, triangulation projects in a pagoda with three naves.

Ph. 21. Lombok, grain silo covered with roofing resting on screeds.

Ph. 22. Jakarta, Indonesia, framework of the canopy of a Chinese house.

Ph. 23. Panamalai, Tamil Nadu, north-east corner of the temple.

Ph. 24. Chiang Mai, barred windows of a pagoda with three naves and three breaks.

Ph. 25. Mandalay, Burma, longitudinal façade of a building on partially buried piles.

Ph. 26. Mandalay, remains of a pagoda sanctuary.

Ph. 27. Siem Reap, window of a guest house of a pagoda located within the enclosure of Angkor Wat.

Ph. 28. Kudus, Java, window of a 19th century house.

Ph. 29. Salé, Burma, relief of the crown of the cornice of the monastery sanctuary.

Ph. 30. Kudus, street in the old part of the town.

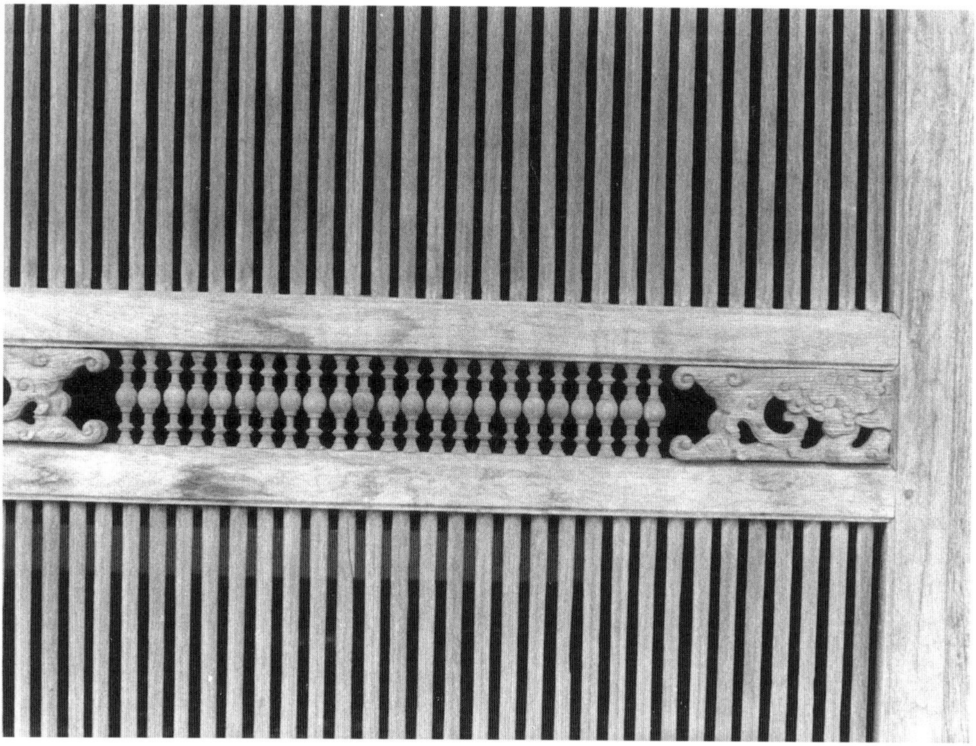

Ph. 31. Kudus, detail of a sliding door.

Ph. 32. Mudianpakhan, Tamil Nadu, west of Madras, brick oven.

Ph. 33. Pagan, Mingalazedi, stupa of the corner of the upper terrace.

Ph. 34. Pagan, barrel vault.

Ph. 35. Sambor Prei Kuk, inner wall of an octagonal tower made of brick, with sandstone hooks supporting the ceiling beams.

Ph. 36. Angkor, entry pavilion of the first terrace of the temple of Ta Kev, corbel covering the northern wing.

Ph. 37. Candi Jabung (eastern Java), eastern façade.

Ph. 38. Principal sanctuary of Padang Lawas (central Sumatra).

Ph. 39. Nindowari, Pakistan. (Photo by J.M. Casal), silo XLII of the silo.

Ph. 40. That i Bahi, Pakistan, inner door of monastery.

Ph. 41. Chivaki, Afghanistan, base of the principal stupa (photo from 1957).

Ph. 42. Kerala, quarry tool.

Ph. 43. Kerala, laterite carrier.

Ph. 44. Gangaikondacholapuram, stereotomy of the enclosure.

Ph. 45. Dieng Candi Bhima, corbel covering the cella.

Ph. 46. Prambanan, temple of Siva, southern wall of the first enclosure.

Ph. 47. Valdauur, Tamil Nadu, lime ovens. The structure of the foreground is an empty oven. In the background is a cooling oven with the fire put out.

Ph. 48. Prasat Kravan (Angkor), central tower, southern internal relief (photo by L. Ionesco).

Ph. 49. Bantay Kdei, overhauling of the second construction step obliterating the southern pediment of the central tower.

Ph. 50 Candi Jago, repair of the faulty layout of the second register of the foundation of the principal sanctuary.

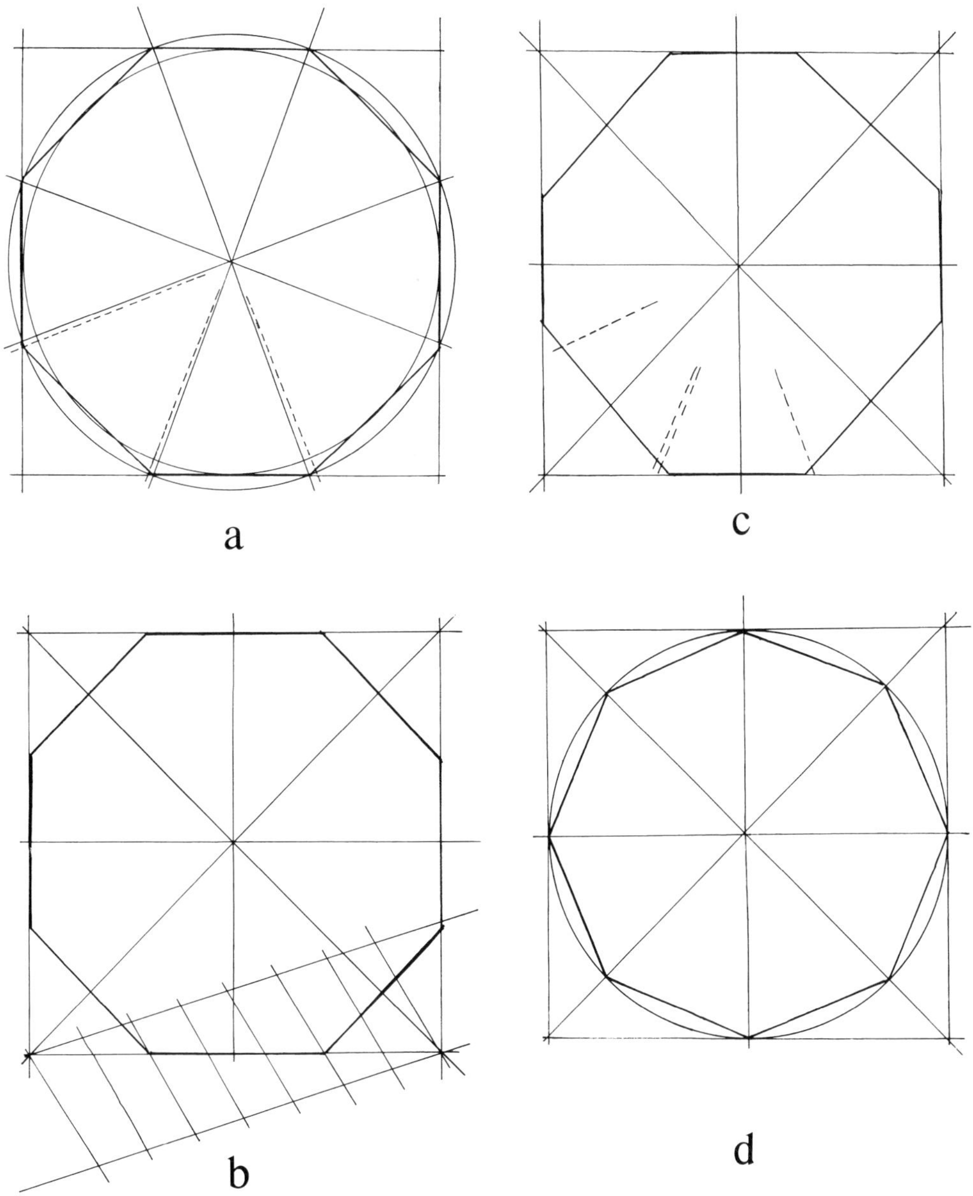

Fig. 1. Construction of an octagon, a, following the Mayamata, the method consisting of adding 1/24th of the side of the initial square, b, the method using 3/7th of the square, c, that using the average between the third and quarter of the square; finally, d, the usual construction in the west. The dotted lines of drawings a and c indicate, for comparative purposes, the exact location of the angles.

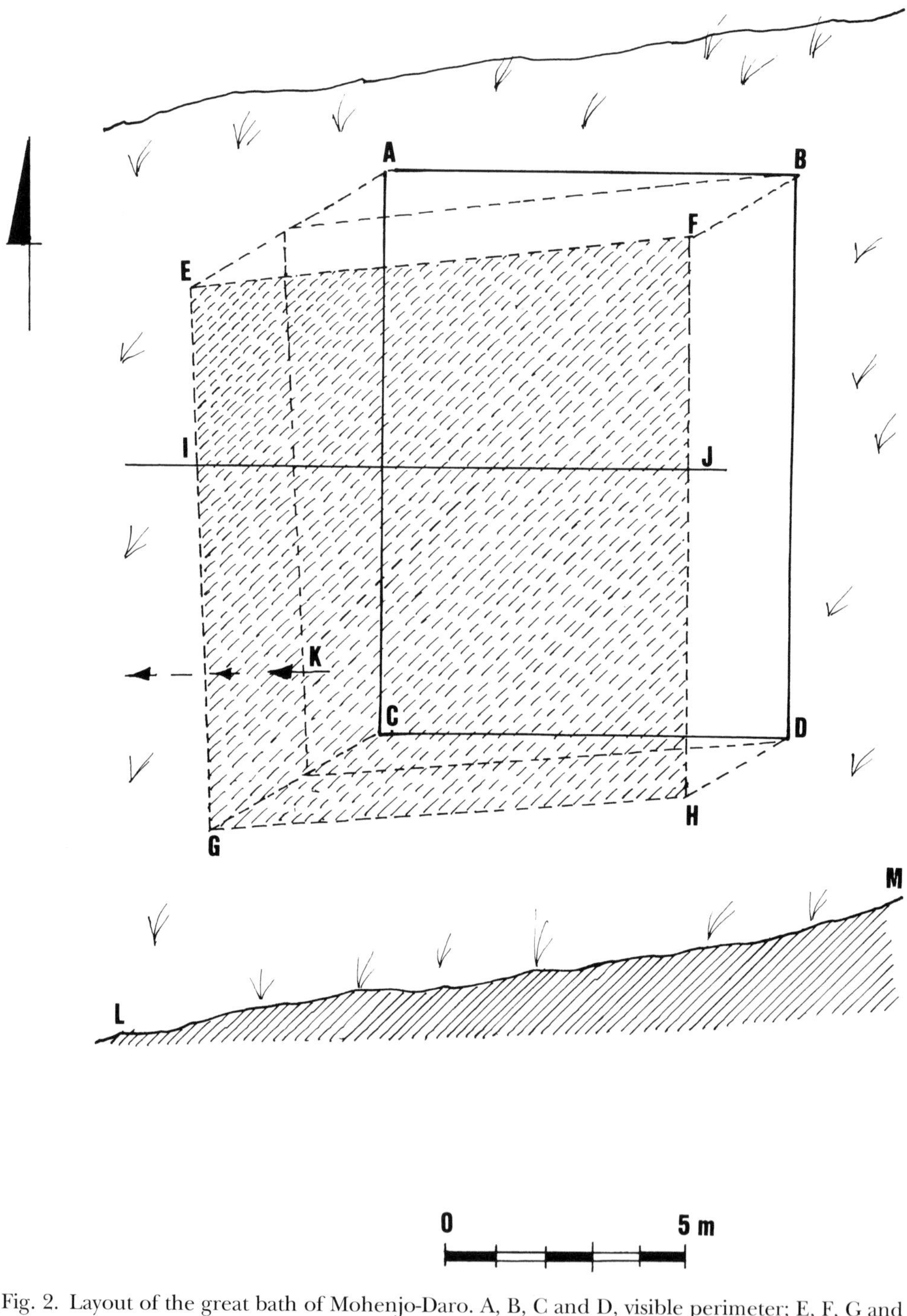

Fig. 2. Layout of the great bath of Mohenjo-Daro. A, B, C and D, visible perimeter; E, F, G and H, reconstituted layout. I and J, linked with the archaeological department. K: water evacuation. M and L, slope of terrain at time of layout.

Fig. 3. Borobudur, fragment of the relief of the fourth gallery, panel 46 of the north side and restitution of the tool carried by the boddhisatva; taking into consideration the scale of the represented characters, the total length of the tool should be one metre.

Fig. 4. Bayon, southern gallery of the first floor, fragment of the relief of the eastern wing.

Fig. 5. Borobudur, plan of the entire group of successive layouts. W, first layout; X, second; Y, third; Z, fourth; V and U, fifth.

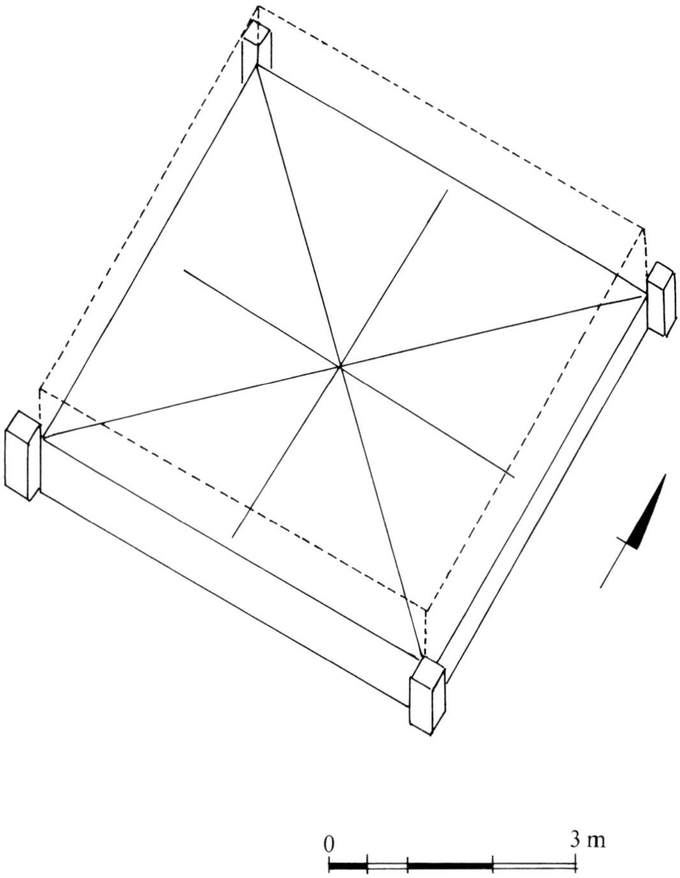

Fig. 6. Candi Sewu, transfer of axes at the level of the ninth course of bricks of the base of the principal statue.

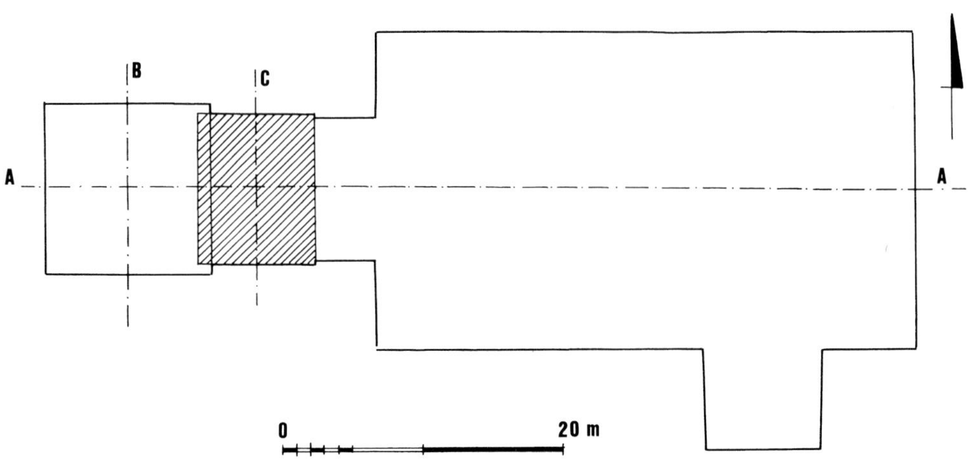

Fig. 7. Darasuram, location of axes A and B of the constructed architecture and C, of the represented architecture.

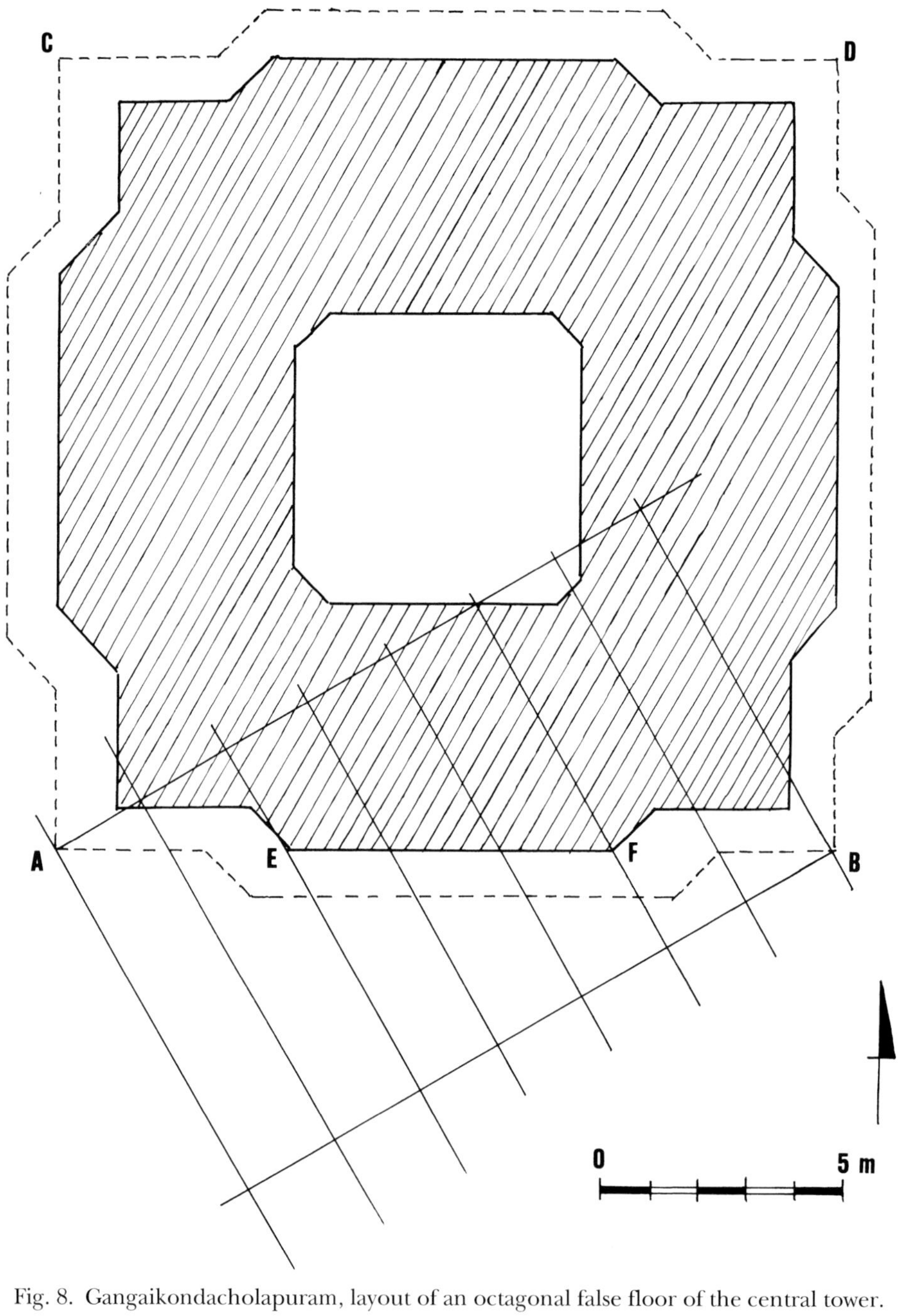

Fig. 8. Gangaikondacholapuram, layout of an octagonal false floor of the central tower.

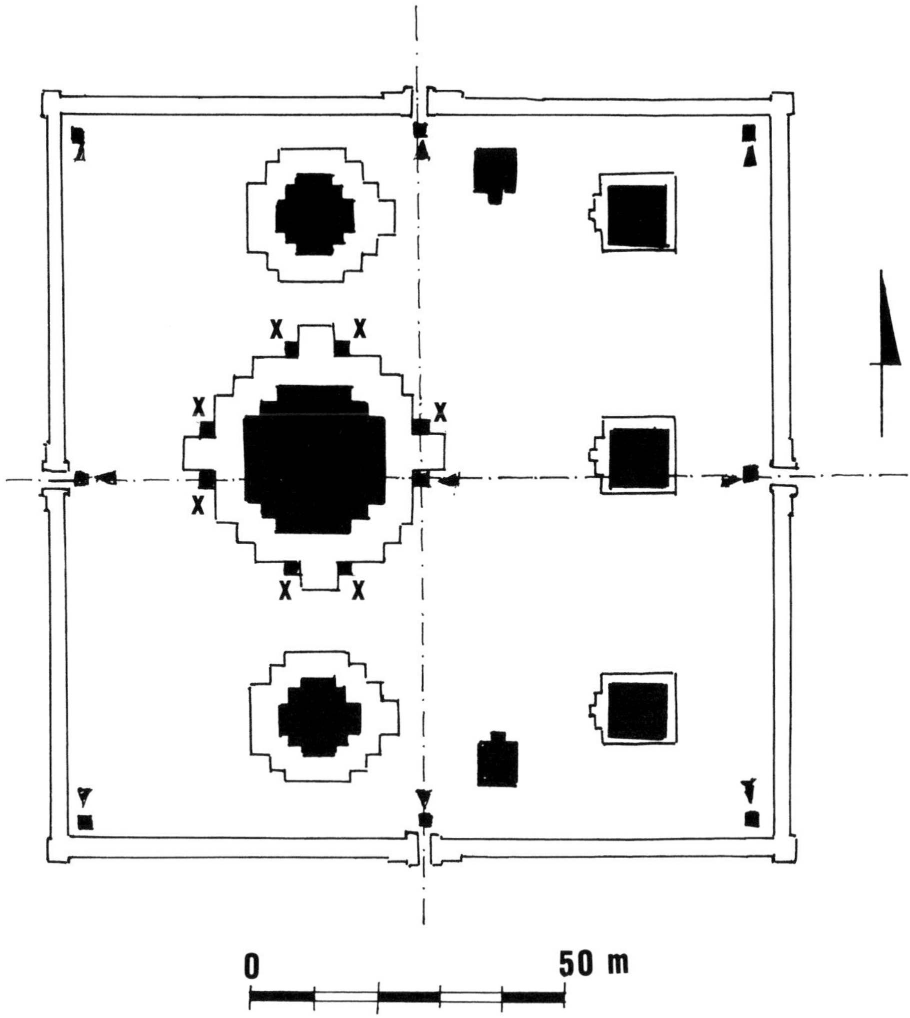

Fig. 9. Prambanan, schematic plan of the upper terrace and layout of the temples housing the bench-marks. The letter X indicates the base temples and the arrows point out the location of the doors.

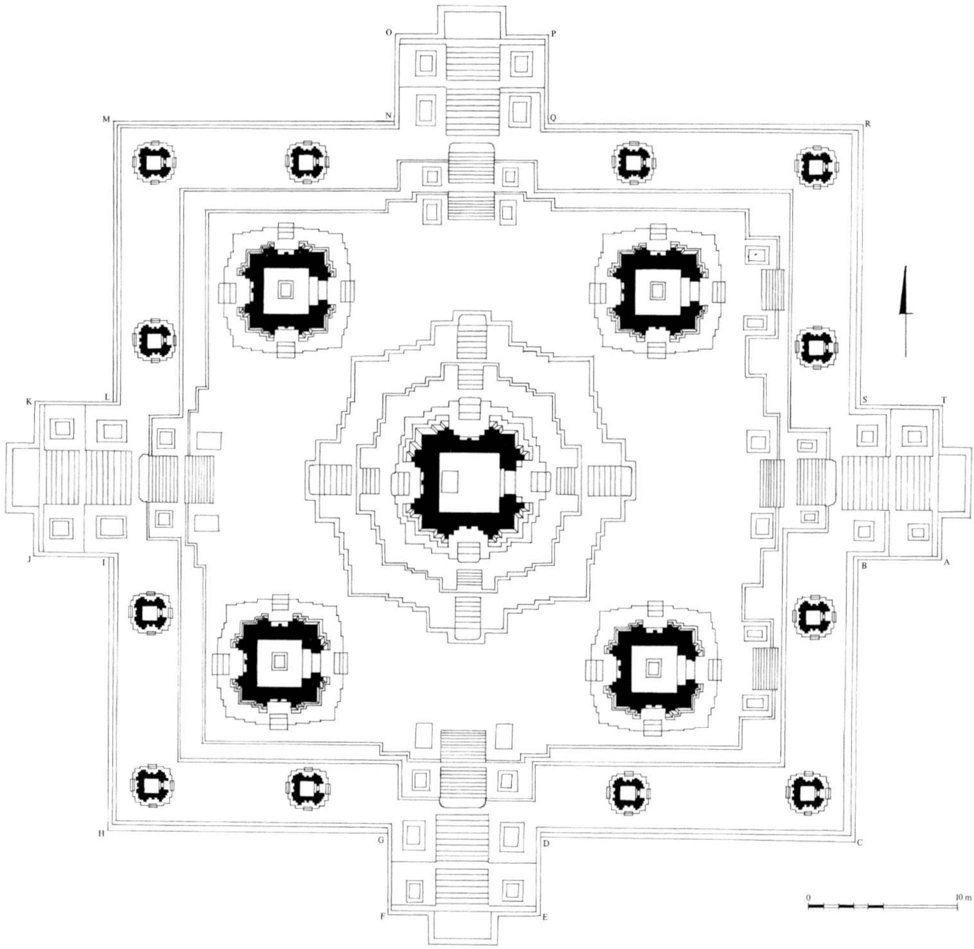

Fig. 10. Pre Rup, plan of the central enclosure of the principal temple. The letters refer to the table on the page.

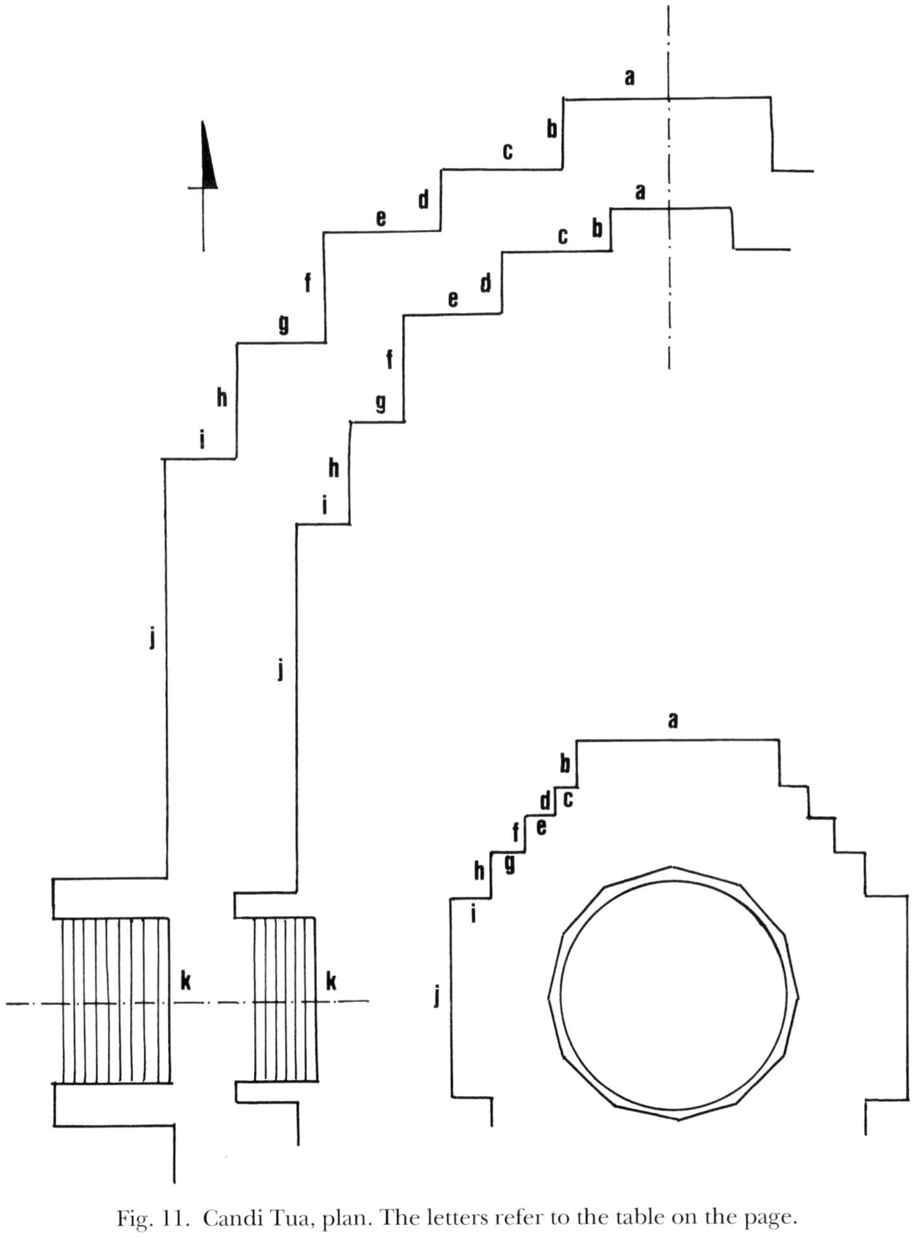

Fig. 11. Candi Tua, plan. The letters refer to the table on the page.

Fig. 12. Fragment of a fresco decorating the Wat Buak Klok Luang at Chiang Mai.

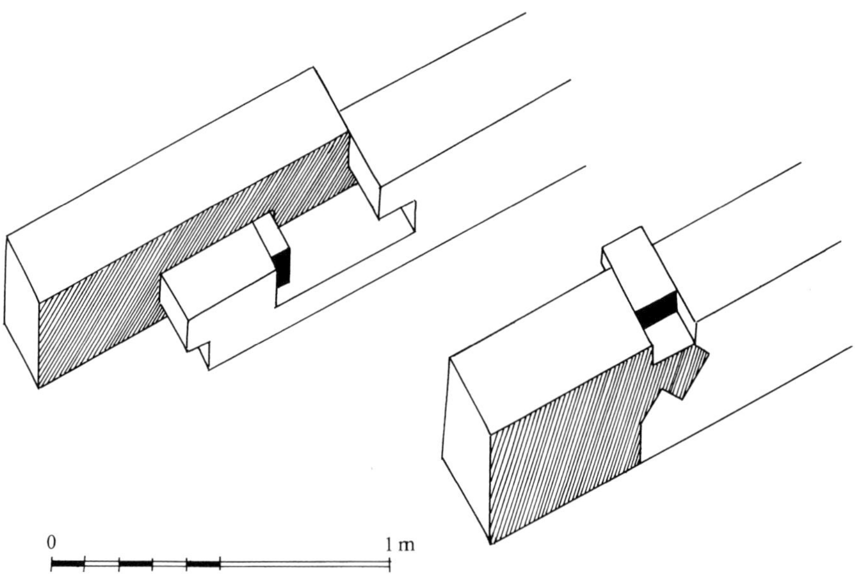

Fig. 13. Assembly of beams in a horizontal plane (line of Jupiter). The elements indicated in black represent the keystones, ensuring consistency.

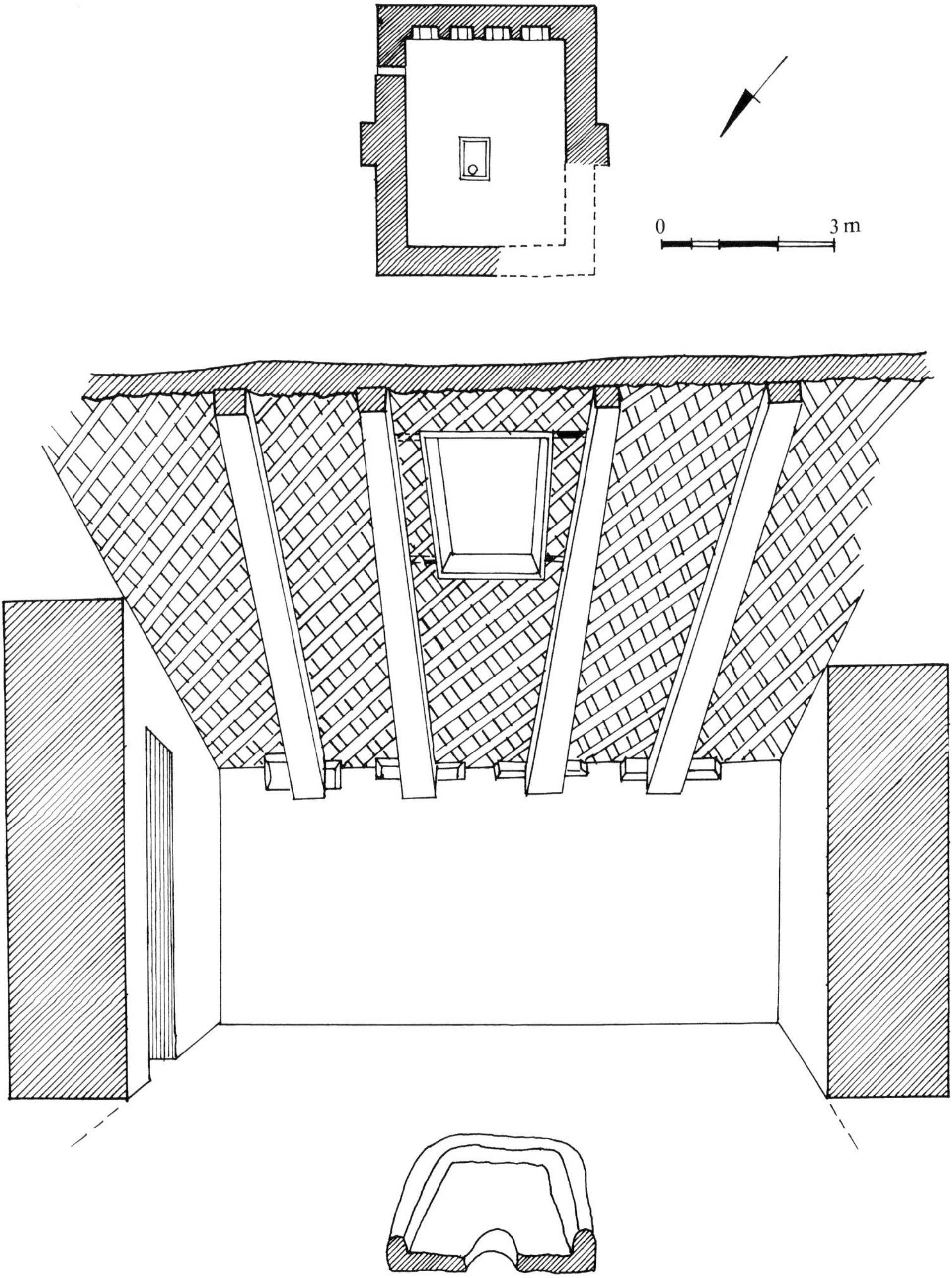

Fig. 14. Mundigak, restitution trial of house CCXXIV, located at level III/2 of the site.

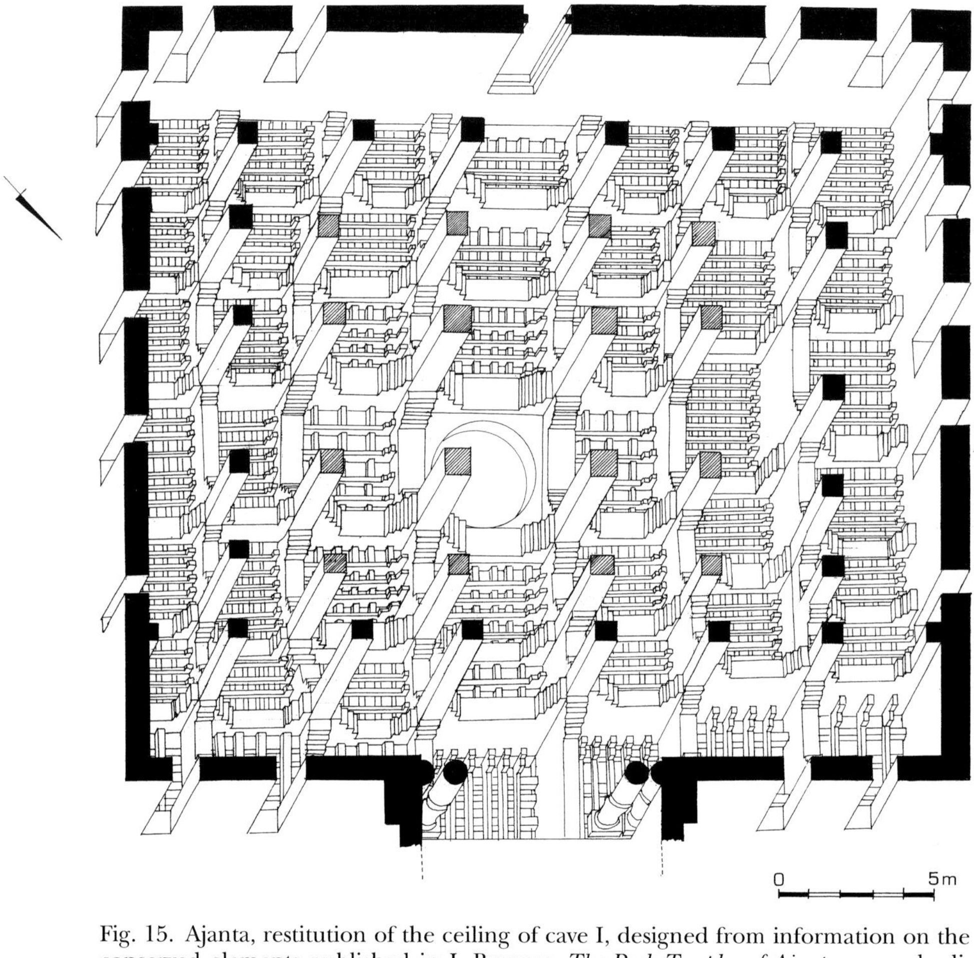

Fig. 15. Ajanta, restitution of the ceiling of cave I, designed from information on the conserved elements published in J. Burgess, *The Rock Temples of Ajanta*, second edition, New York, 1970, pl. II and photographs by Goloubev, op. cit.

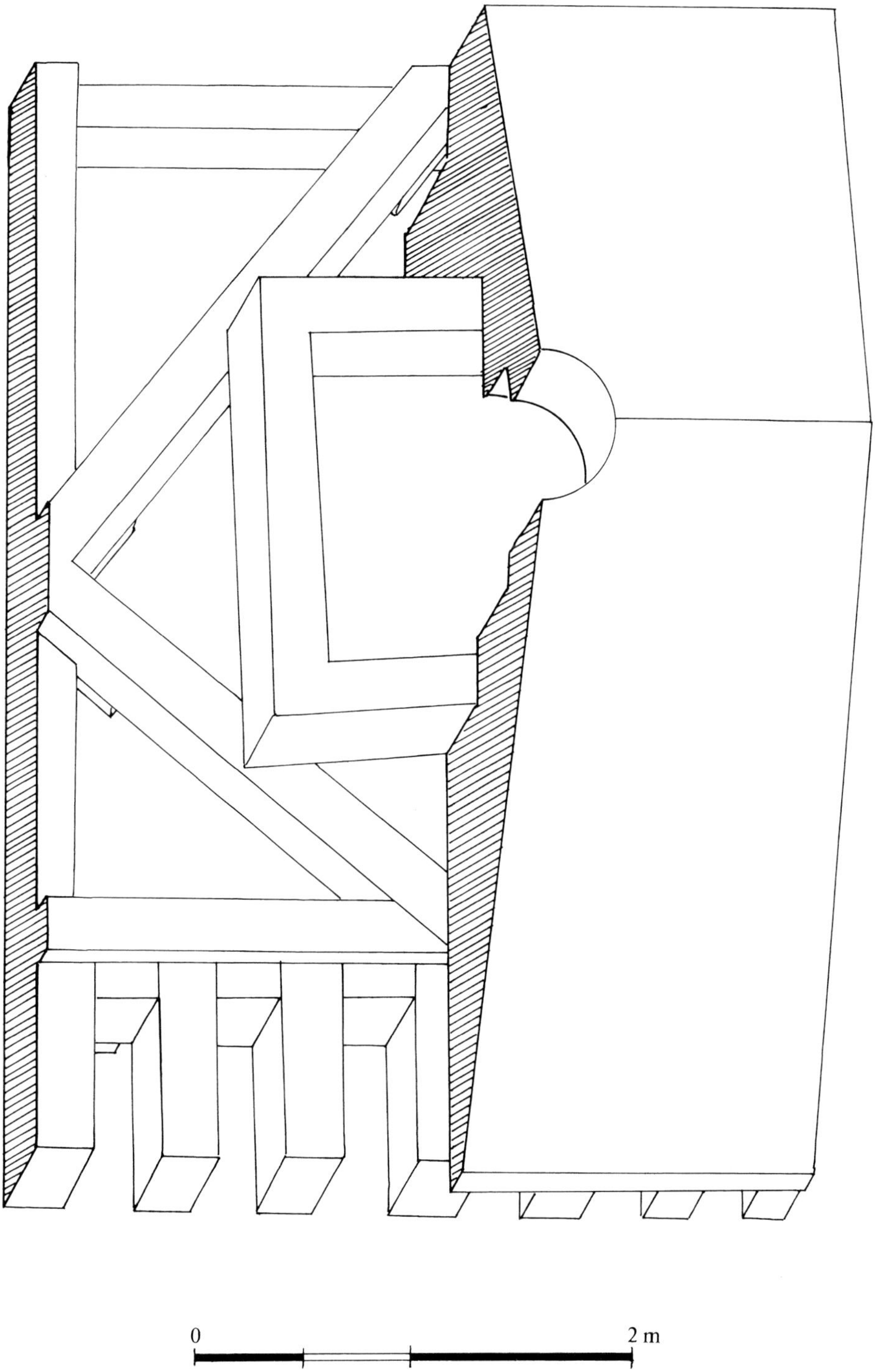

Fig. 16. Foladi, restitution of the roofing of cave C.

Fig. 17. Bamyan, ceiling of cave XV.

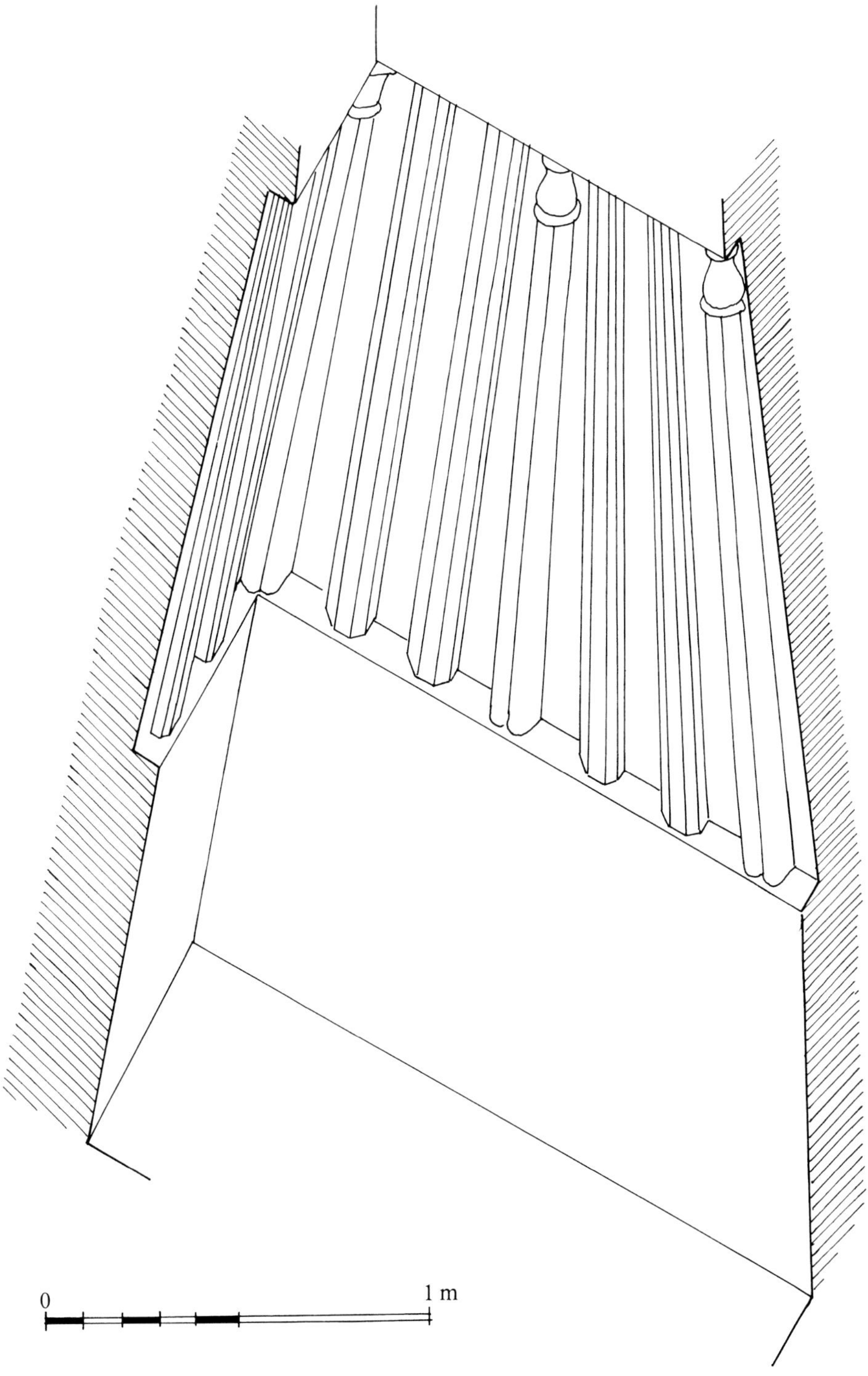

Fig. 18. Bamyan, cave XV, axonometry of the walls.

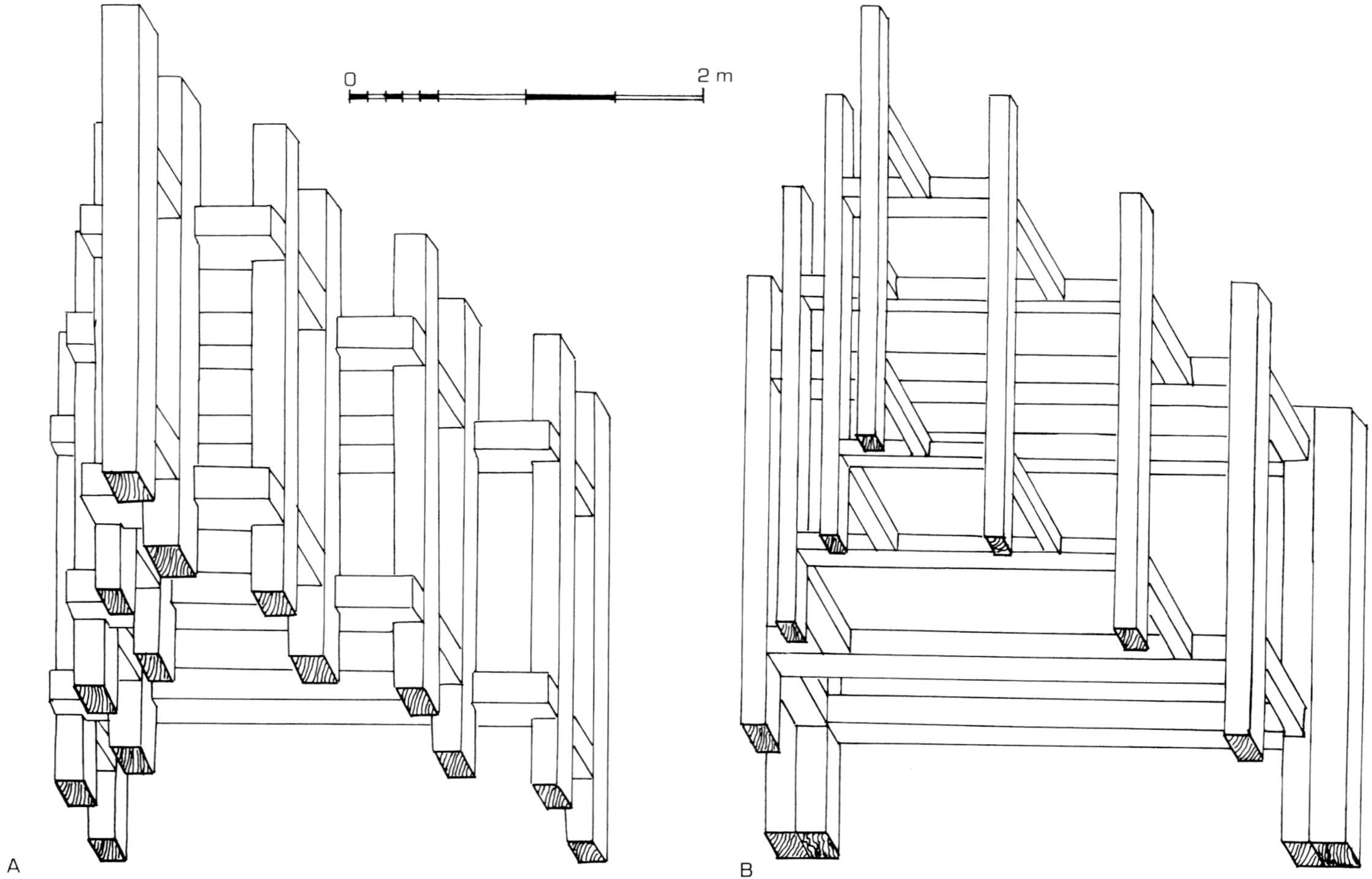

Fig. 19. Bending frames: A, frame supporting the roofing of a porch at Ta Prohm in Angkor; B, frame supporting the roofing of a pagoda of Siem Reap in Cambodia.

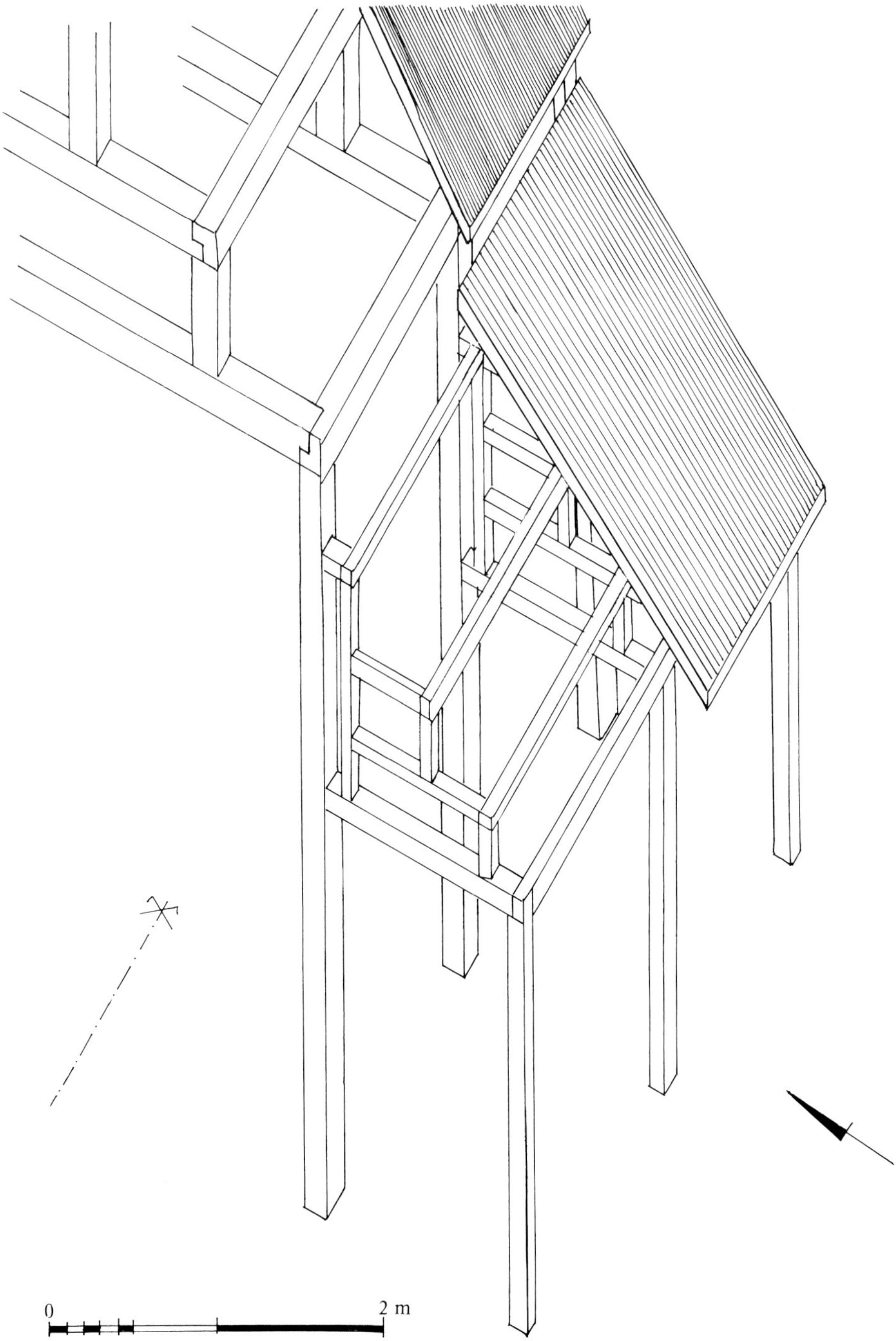

Fig. 20. Bending frame of the side naves of the pagodas of the region of Chiang Mai.

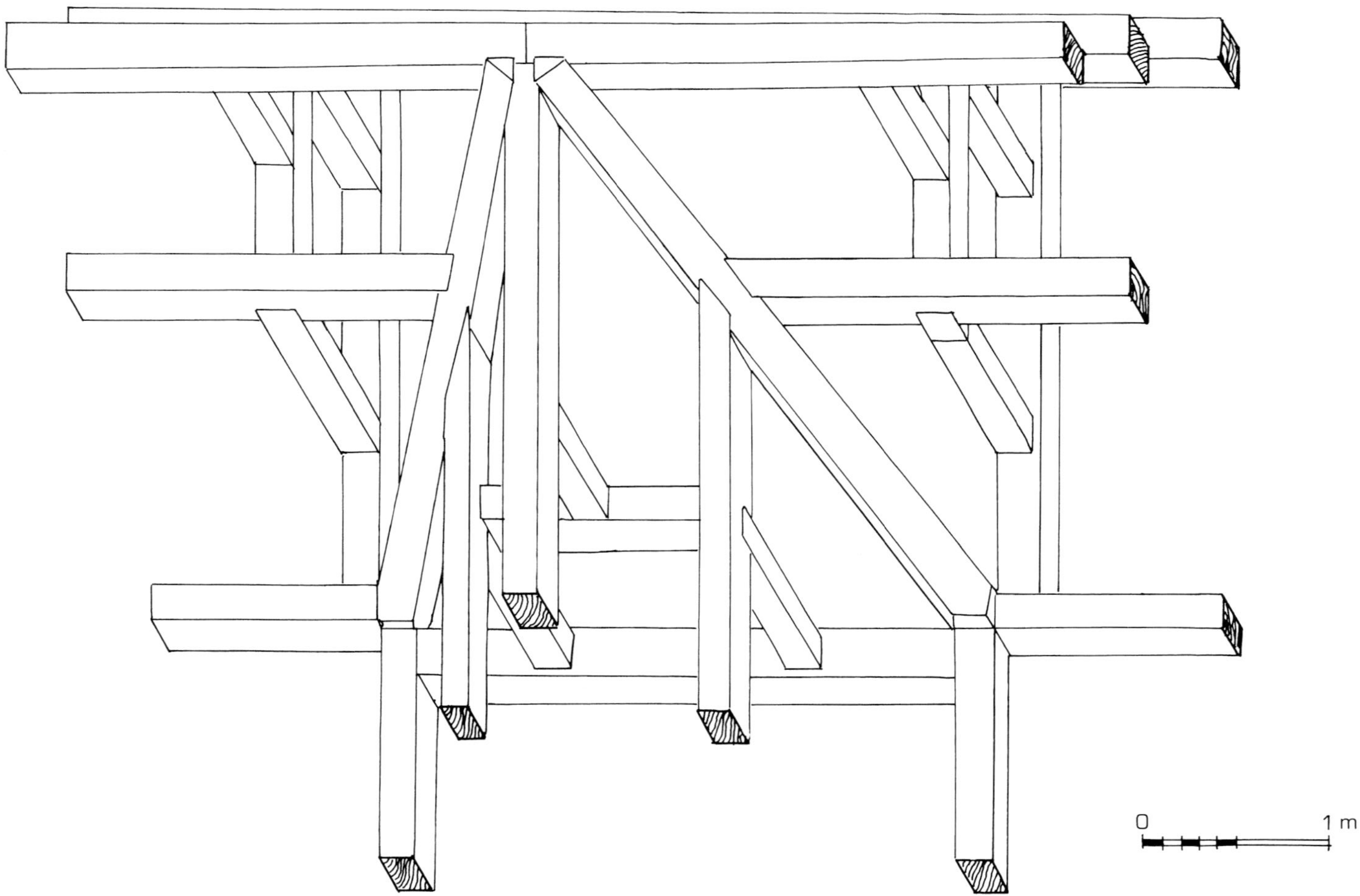

Fig. 21. Bending frame with penetration.

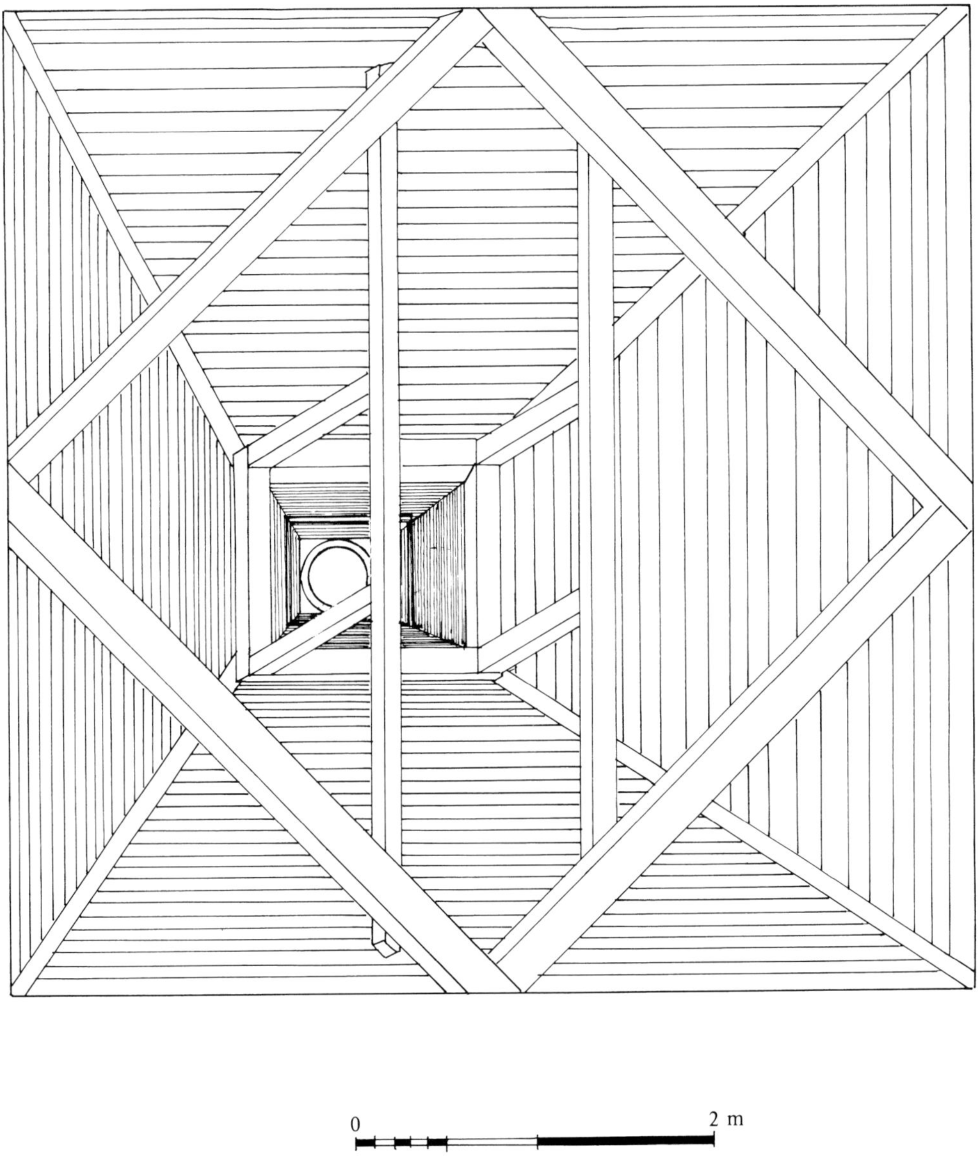

Fig. 22. Bending frame on a square plane, panelled (pavilion of the lacquer Buddha at Salé, Burma).

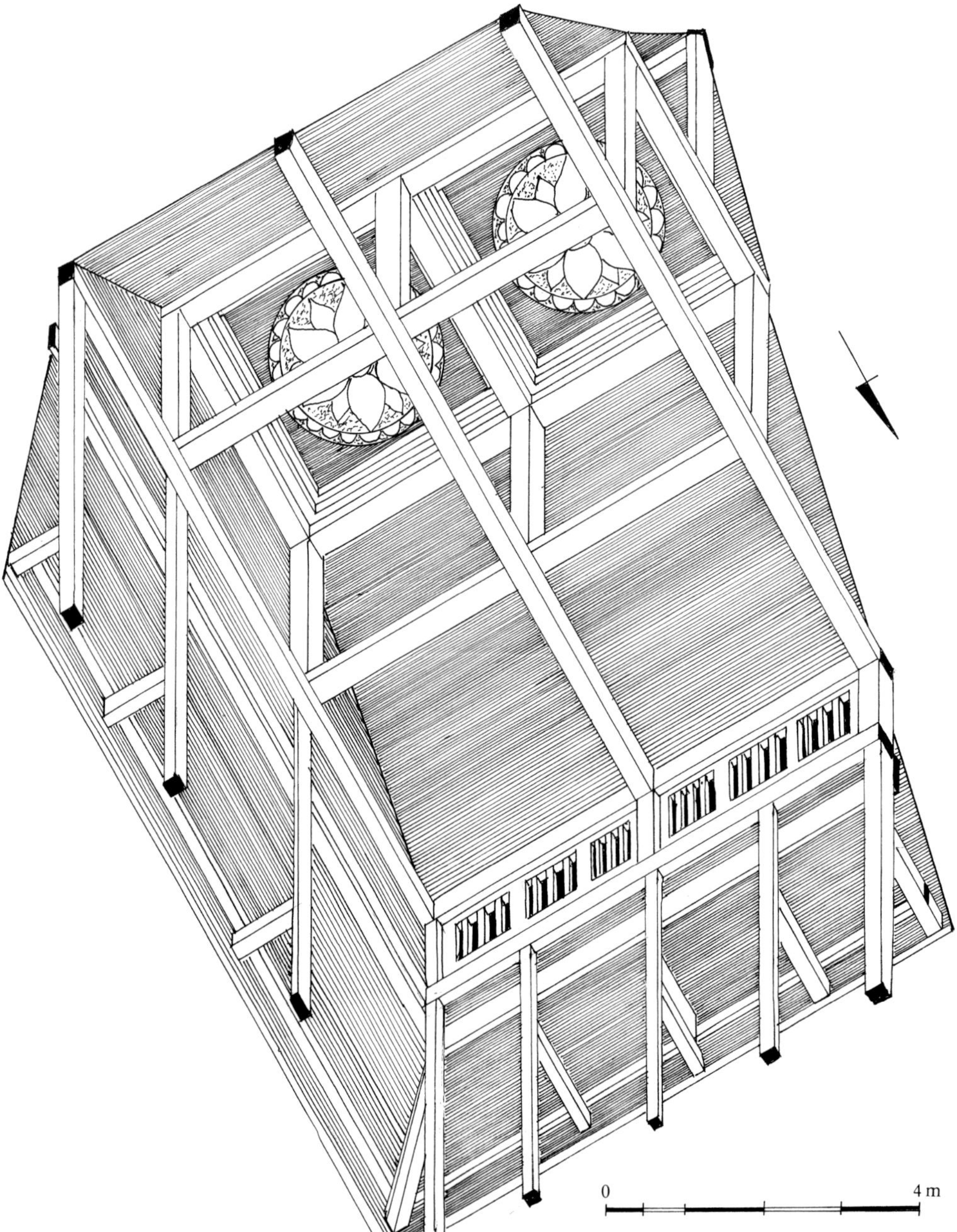

Fig. 23. Bending frame with partial ceiling and panelling, in Amarapura, Burma.

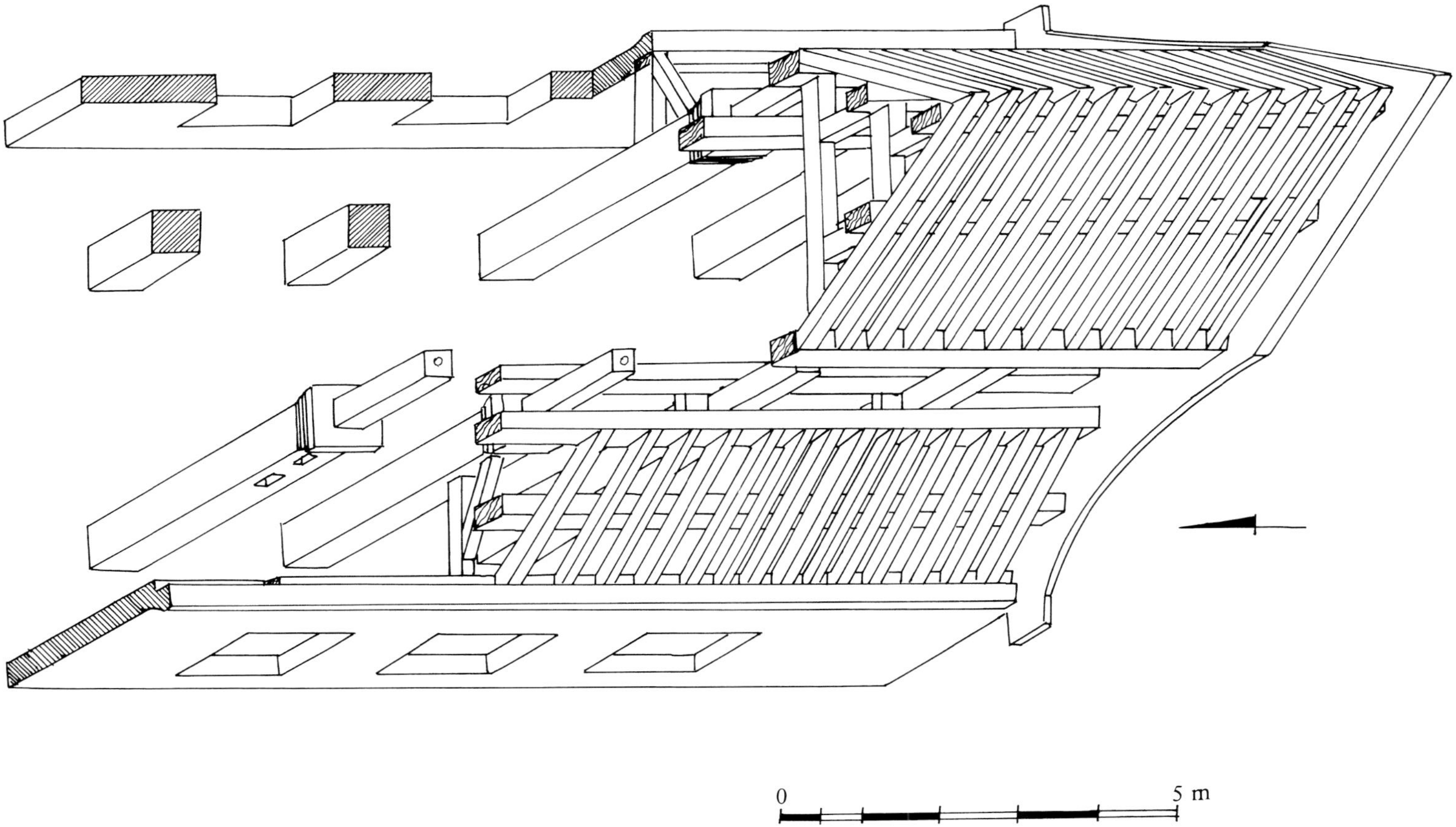

Fig. 24. Preah Vihear, Cambodia, axonometry of room N.

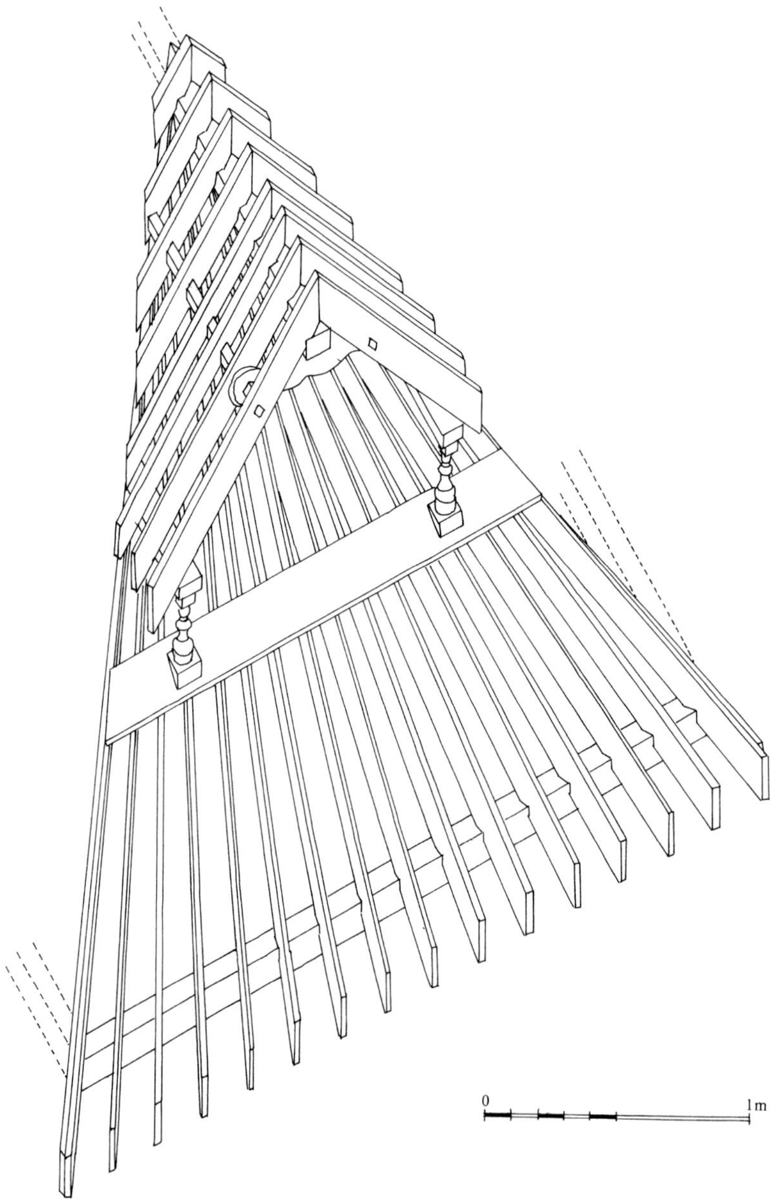

Fig. 25. Implementation of the frame of a false gable over a building covered with a radiating frame (Battambang region).

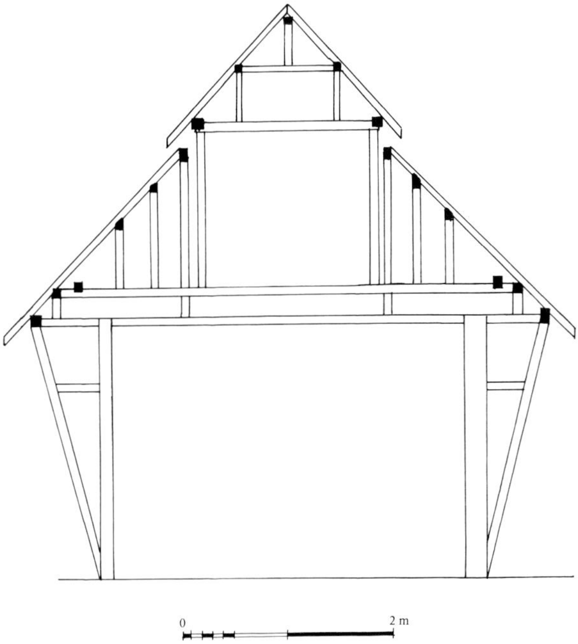

Fig. 26. Fragment of a bending frame with two sides over a structure with warped walls at Luang Prabang in Laos.

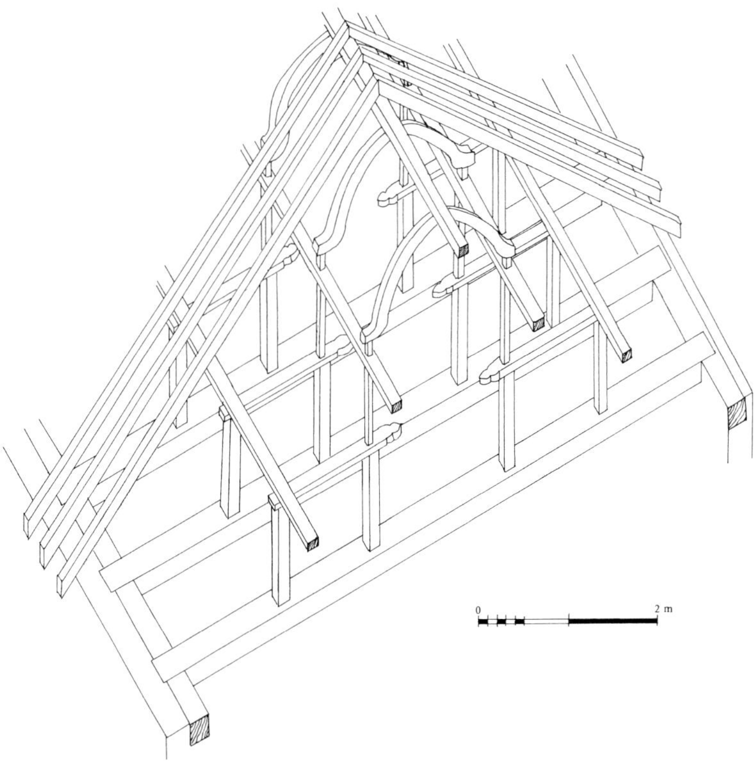

Fig. 27. Bending frame without crown post in the Darasuram region, India.

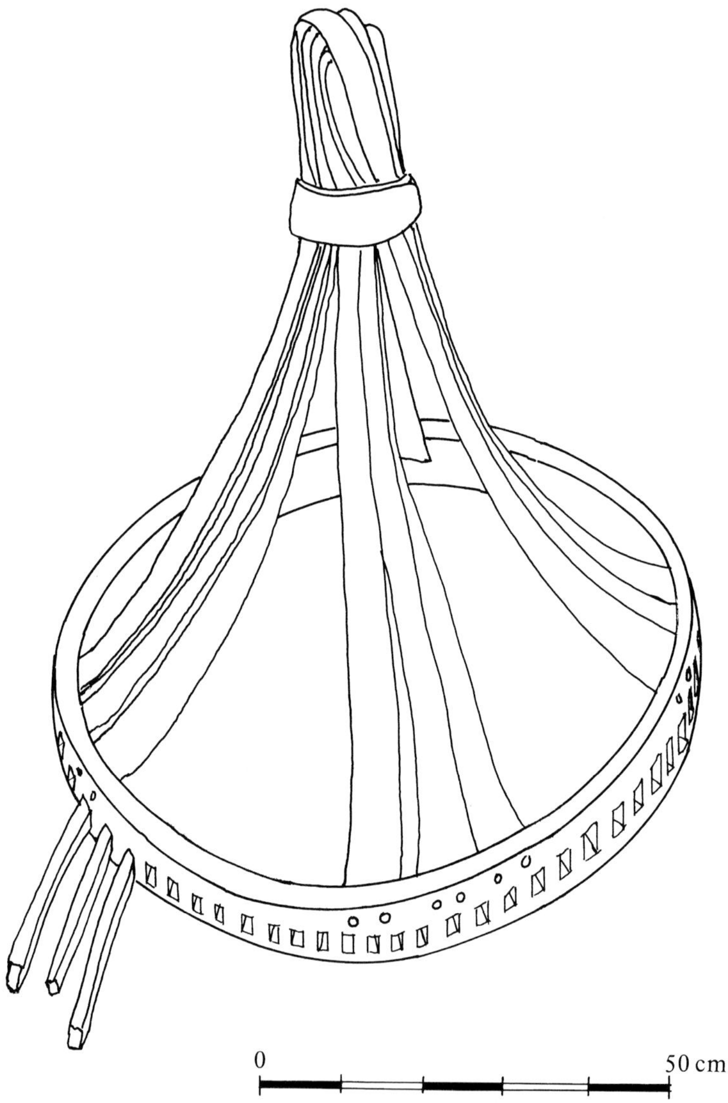

Fig. 28. Ridge finial of a yurt.

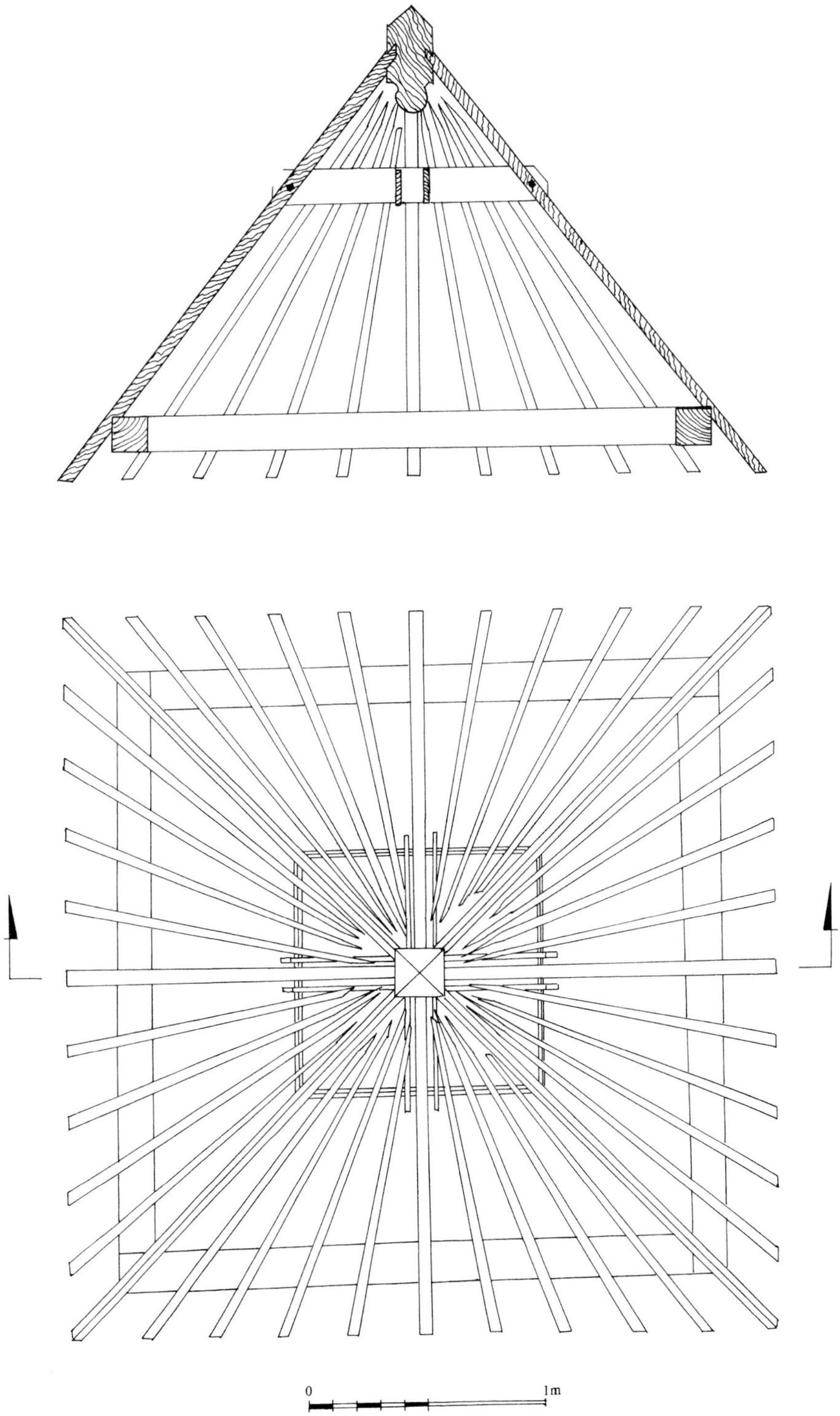

Fig. 29. Radiating frame of Kerala.

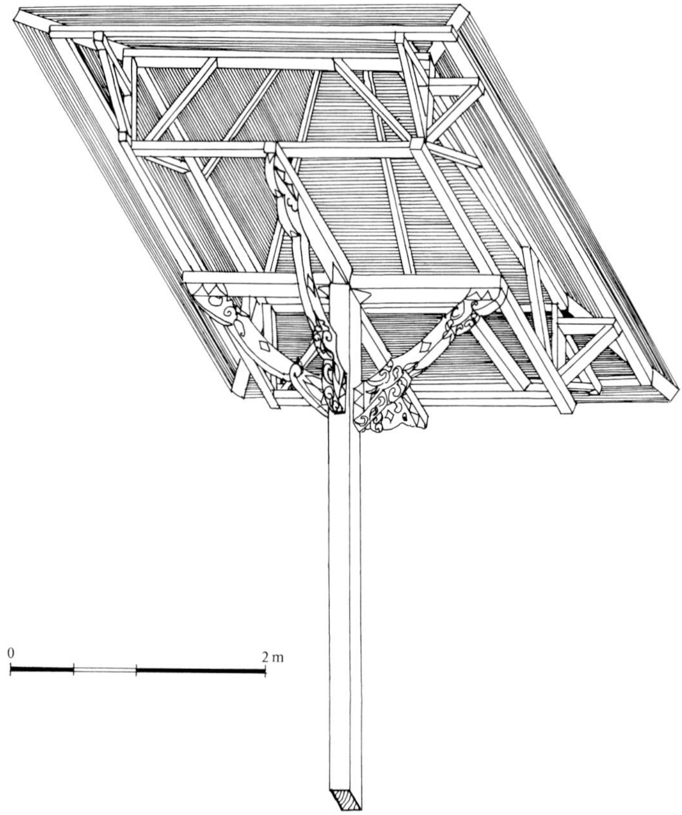

Fig. 30. Radiating frame over a central pillar in Java.

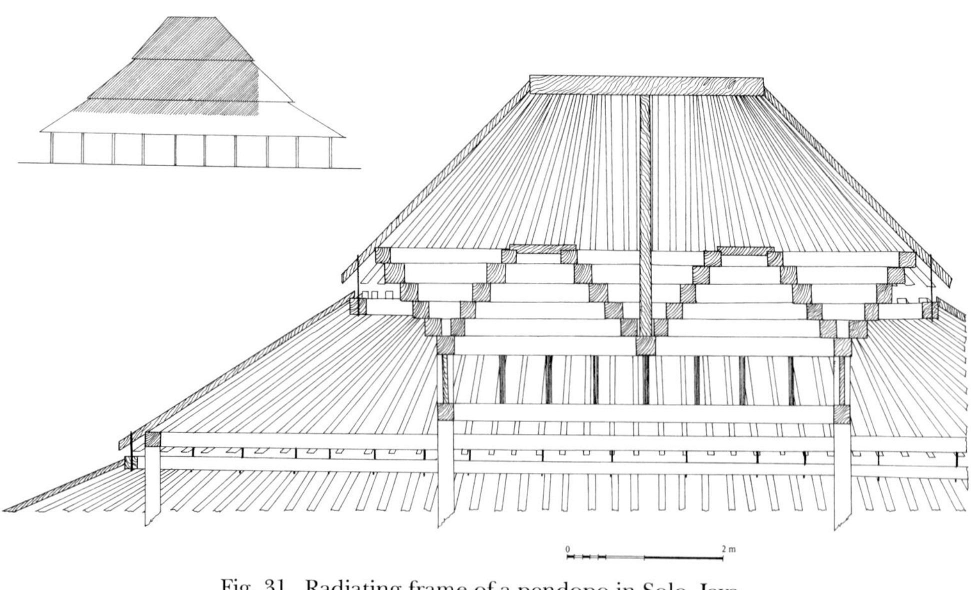

Fig. 31. Radiating frame of a pendopo in Solo, Java.

Fig. 32. Pediment of the temple of Tiruttani, India.

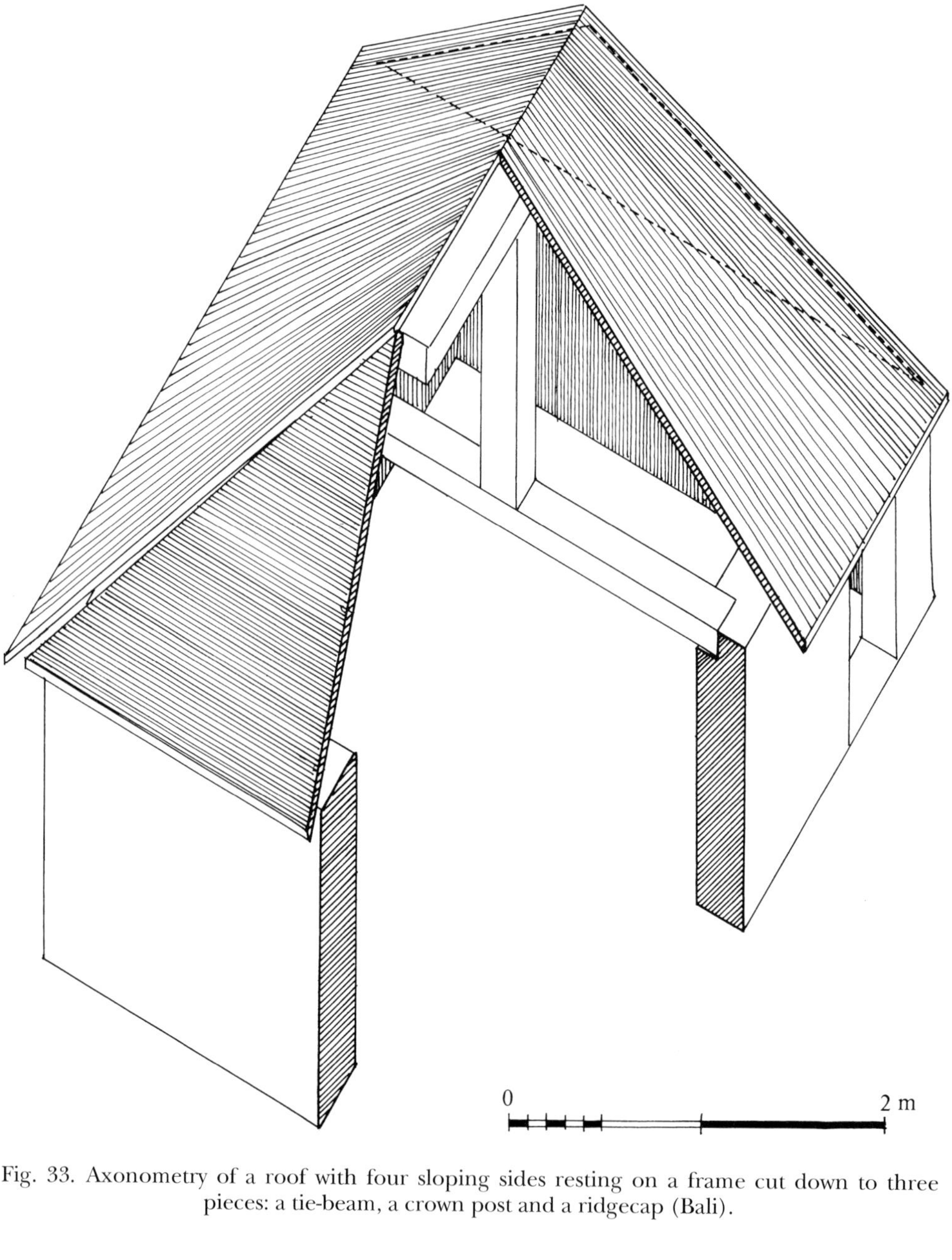

Fig. 33. Axonometry of a roof with four sloping sides resting on a frame cut down to three pieces: a tie-beam, a crown post and a ridgecap (Bali).

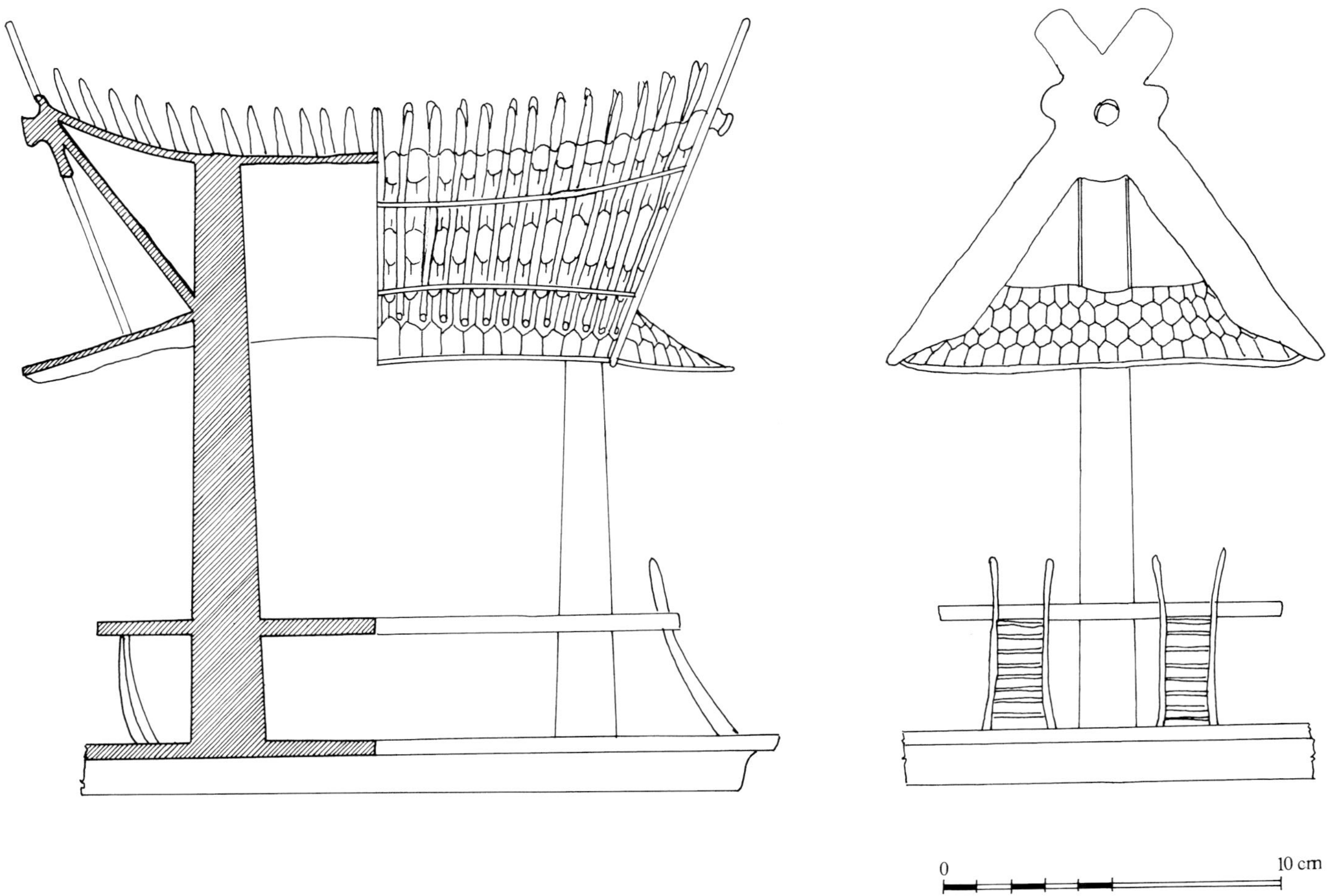

Fig. 34. Frame with ridgecap in tension represented on a bronze discovered at Yunnan (Dian excavation).

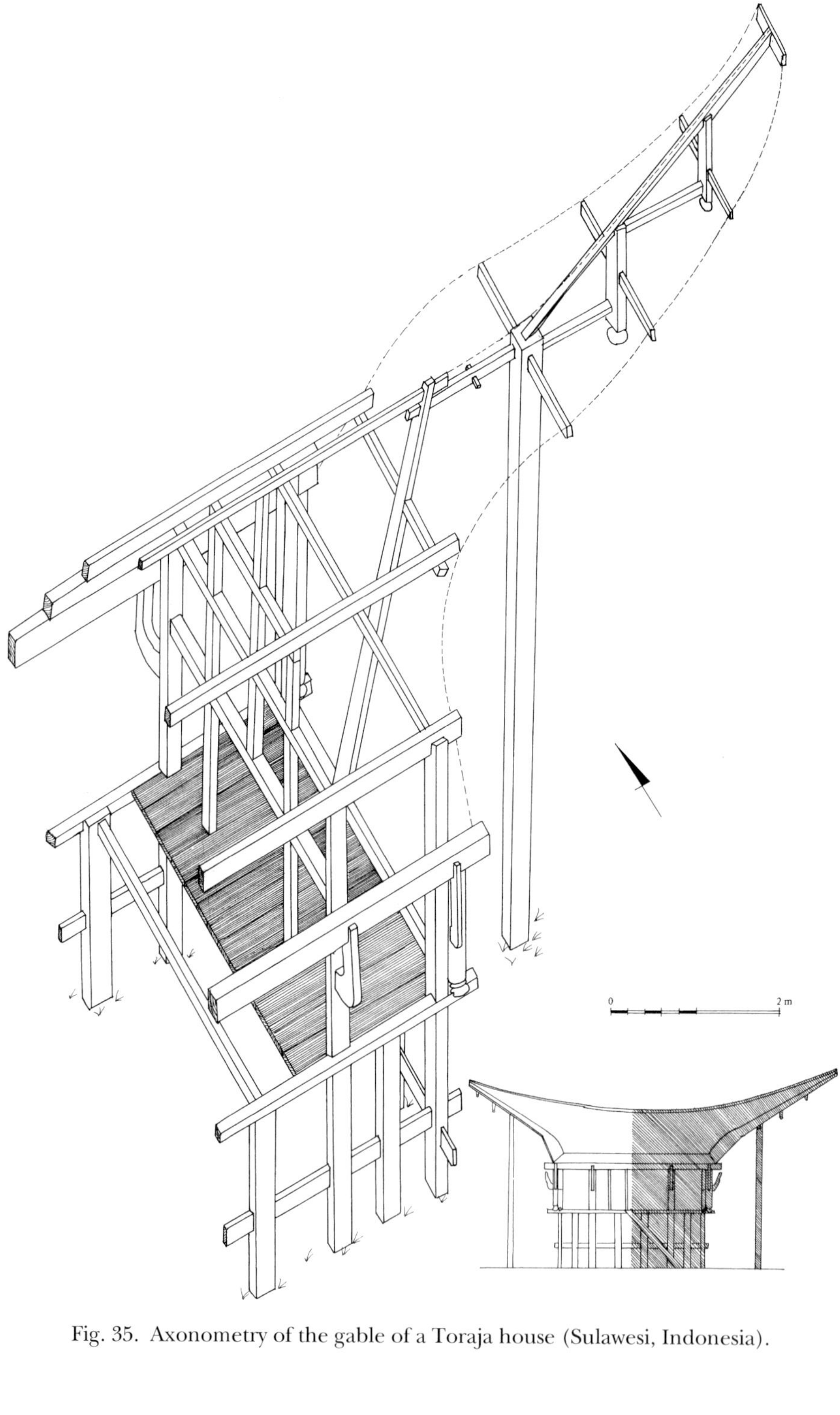

Fig. 35. Axonometry of the gable of a Toraja house (Sulawesi, Indonesia).

Fig. 36. Fragment of a relief discovered in Amaravati (Madras museum).

Fig. 37. Borobudur, represented architecture; the door is shown without a leaf.

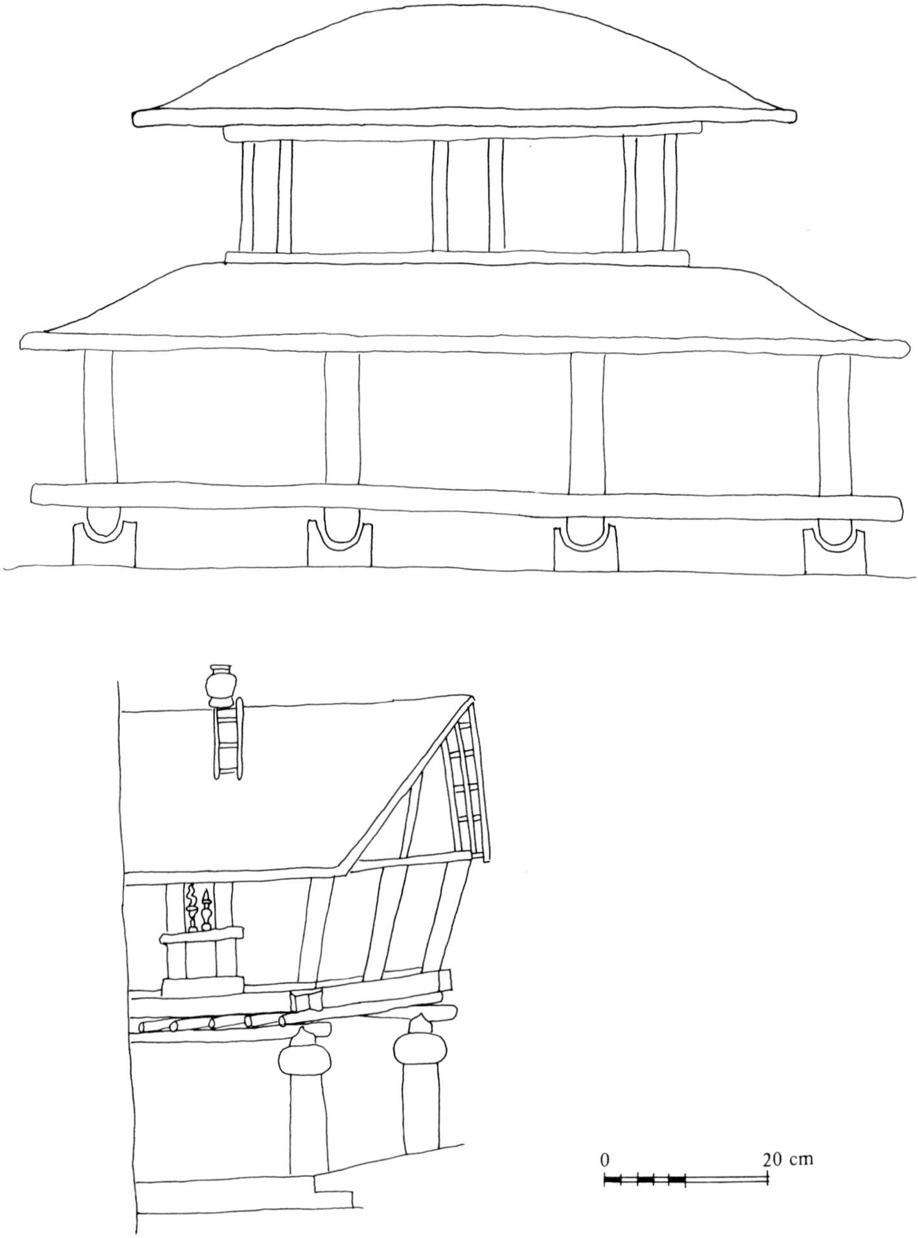

Fig. 38. Borobudur, represented architecture A/ relief O/1/54, b. B/ relief N/1/86, b.

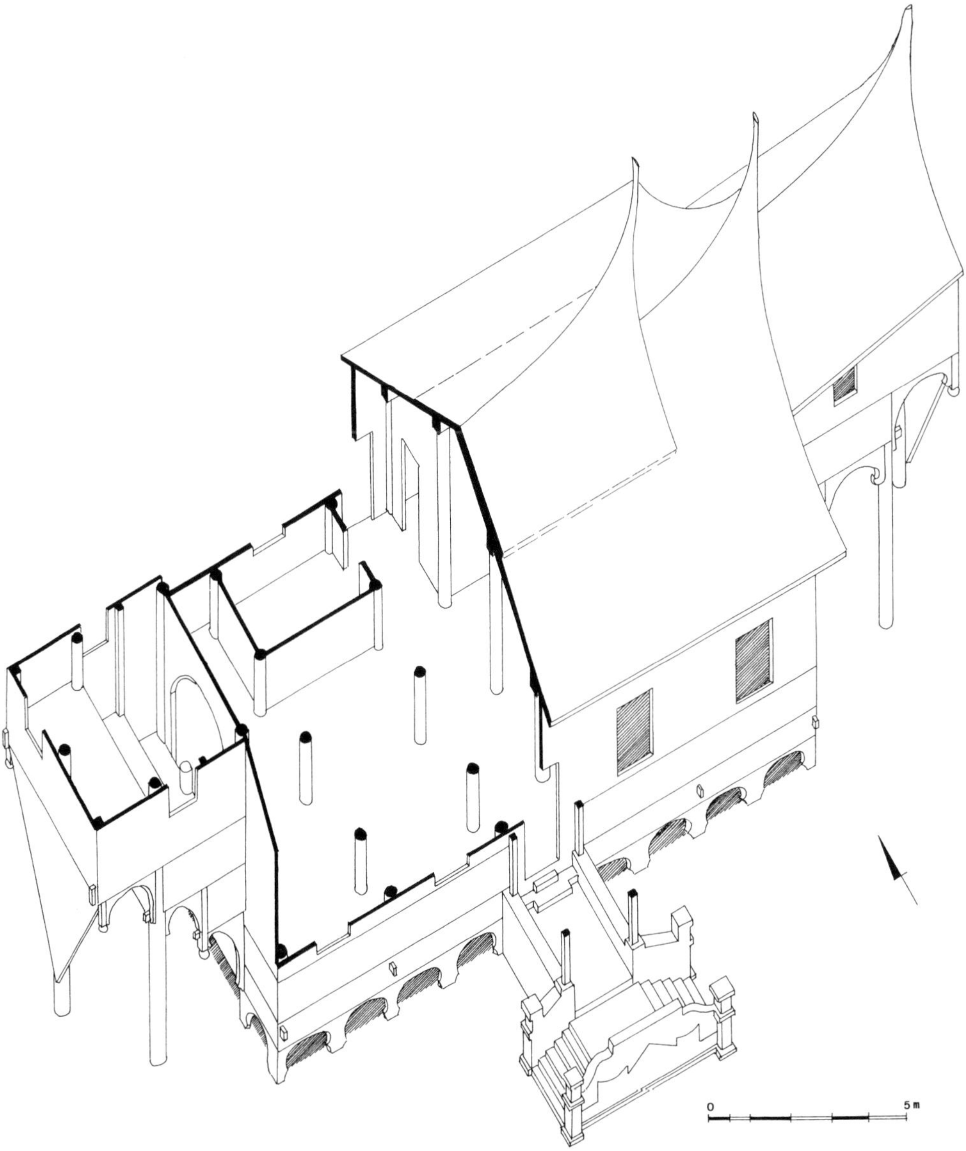

Fig. 39. Minangkabau house, axonometry, Sumatra, Indonesia.

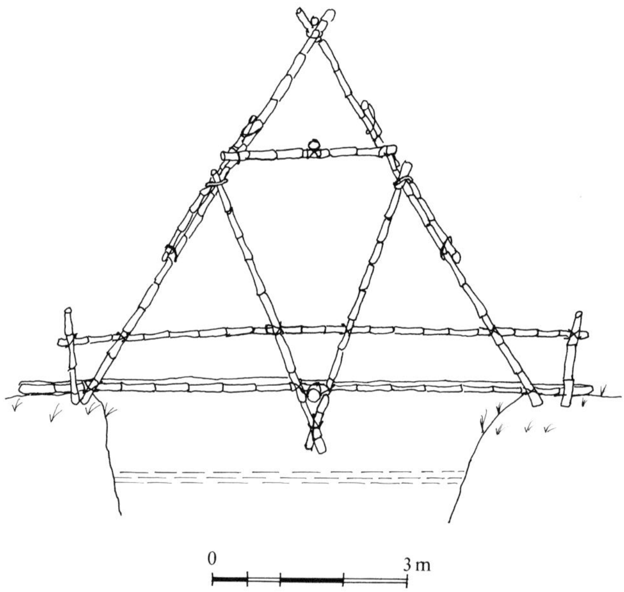

Fig. 40. Bamboo bridge over the river Kanta, Siem Reap region.

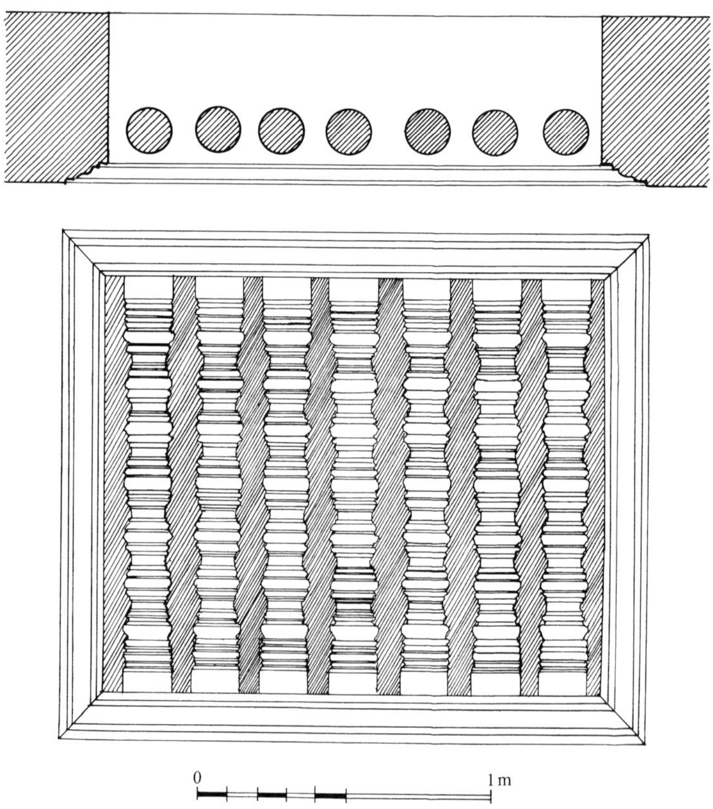

Fig. 41. Sandstone window of the fourth floor of Angkor Wat, outer side.

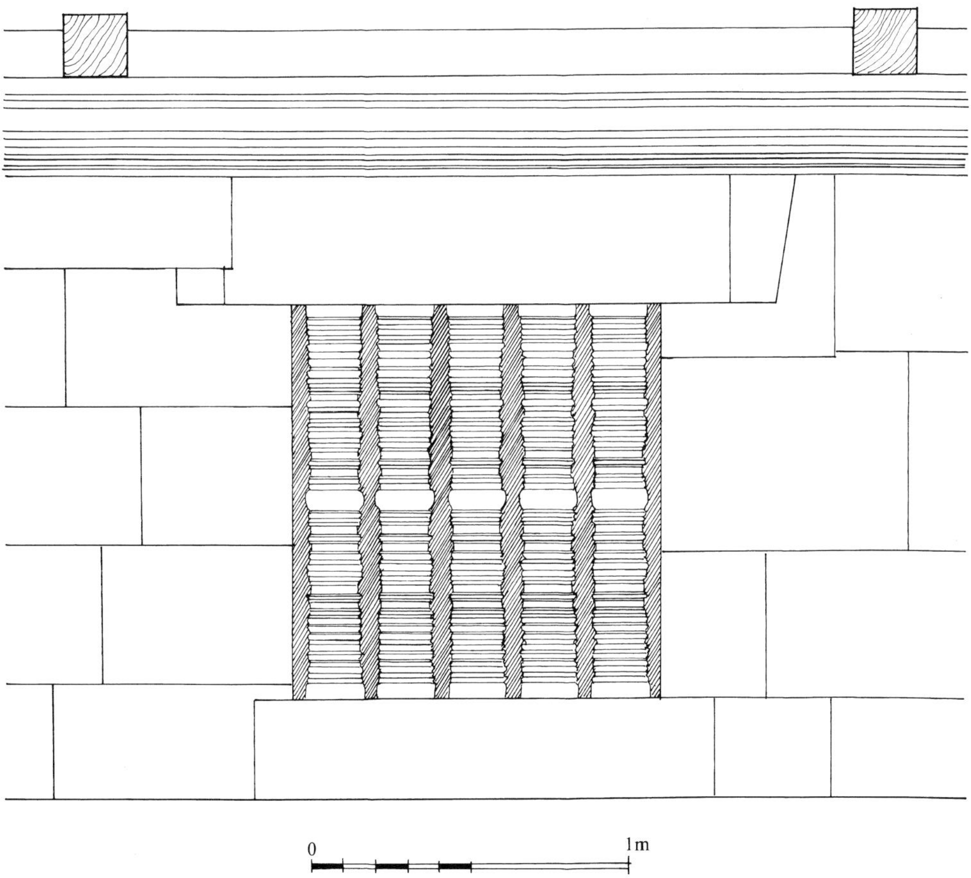

Fig. 42. Sandstone window of room N of Preah Vihear, inner side.

Fig. 43. False window of Bantay Kdei.

Fig. 44. Wall of a Malaysian house of Seramban.

Fig. 45. Angkor Wat, false door with leaf, sculpted on the pavilion of the western entry of the exterior enclosure.

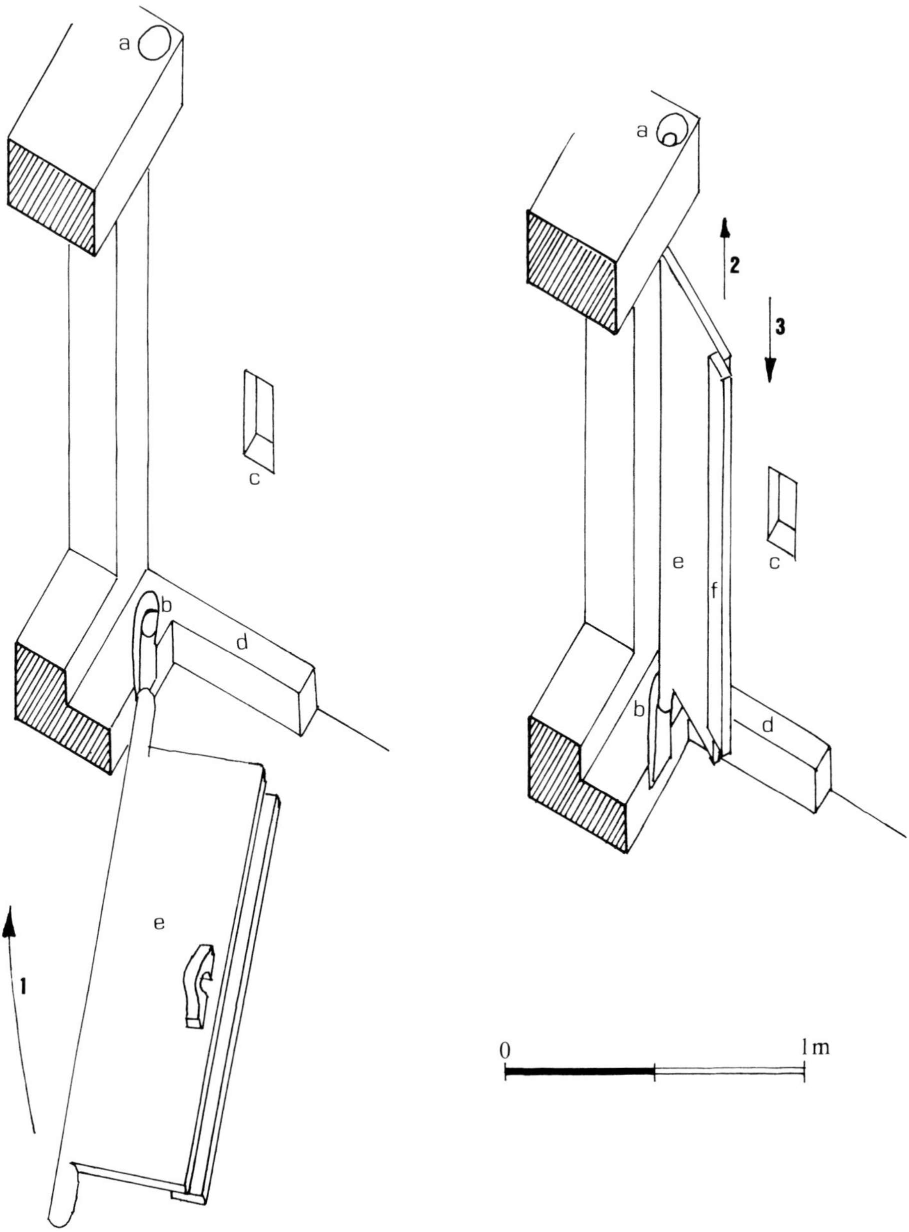

Fig. 46. Implementation of a wooden door leaf in a sandstone sanctuary dating from the 9th century. a/ pivot stone, b/ casement of the socket of the door pivot, c/ casement of the lock, the door being open, d/ support of the leaf, the door being open, e/ closing stile. Arrow 1 indicates the movement of the leaf after the presentation of the axis in the casement of the socket of the door pivot. Arrow 2 indicates the setting of the axis in the pivot stone during the putting into place of the socket of the door pivot. Arrow 3 indicates the setting of the axis in the socket of the door pivot.

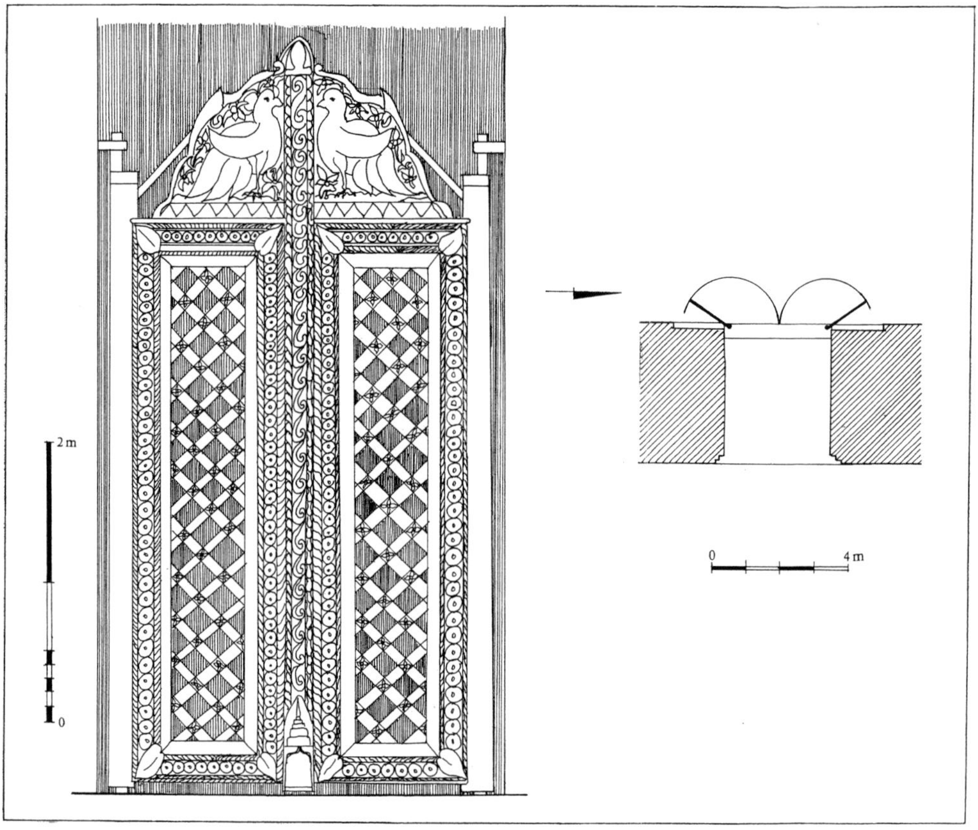

Fig. 47. Pagan, temple of Ananda, western interior door, plan and elevation.

Fig. 48. Pagan, fragment of a fresco of the temple of Ananda Okkaung.

Fig. 49. Ceylon, one-leaf door of a pagoda in the surrounding area of Kandy.

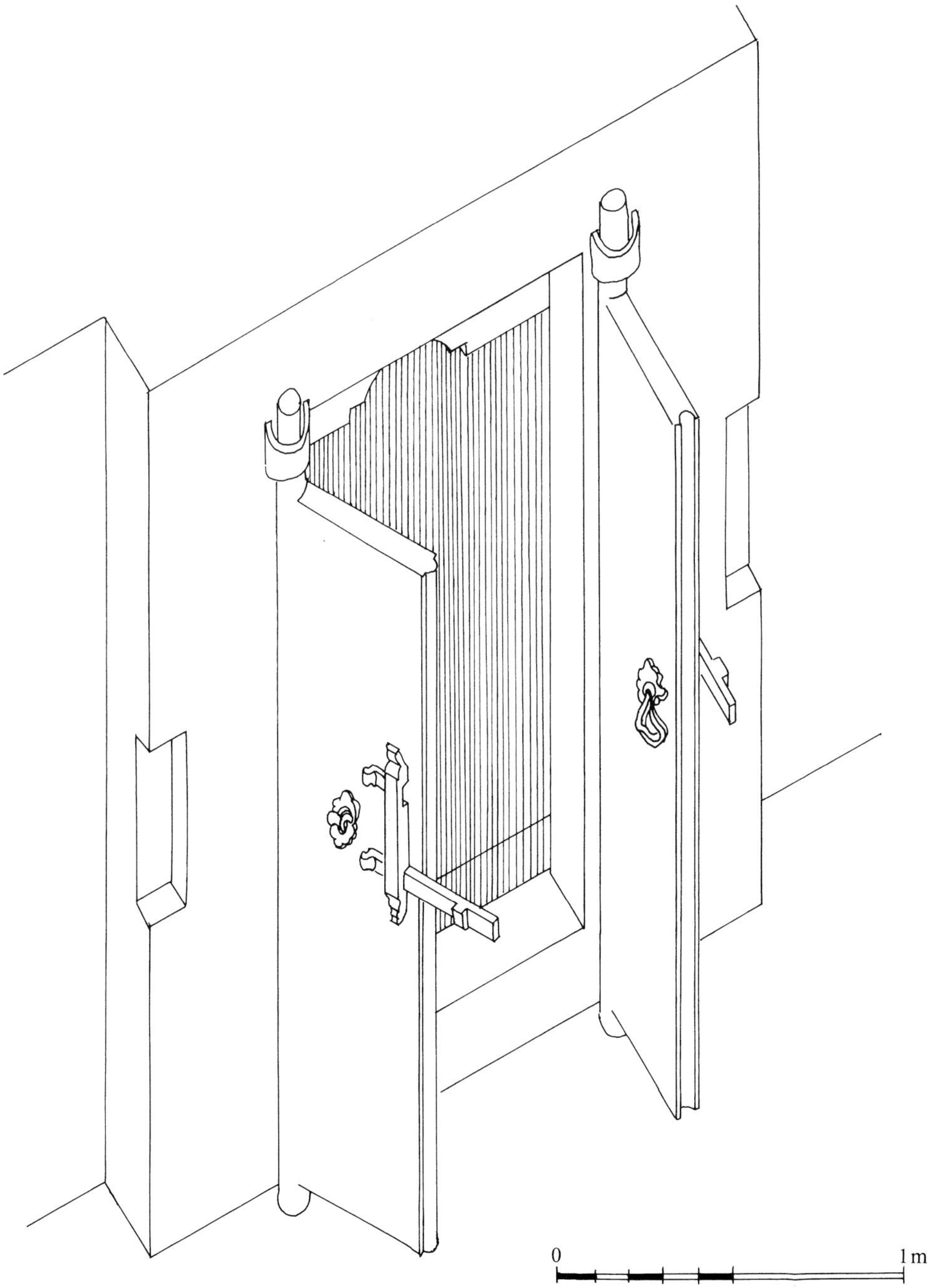

Fig. 50. Ceylon, interior view of a door of the palace of Kandy and its setting in the masonry. A complete door, similar but with a wooden door frame, is on display at the museum of Kandy.

Fig. 51. Luang Prabang, door of pagoda.

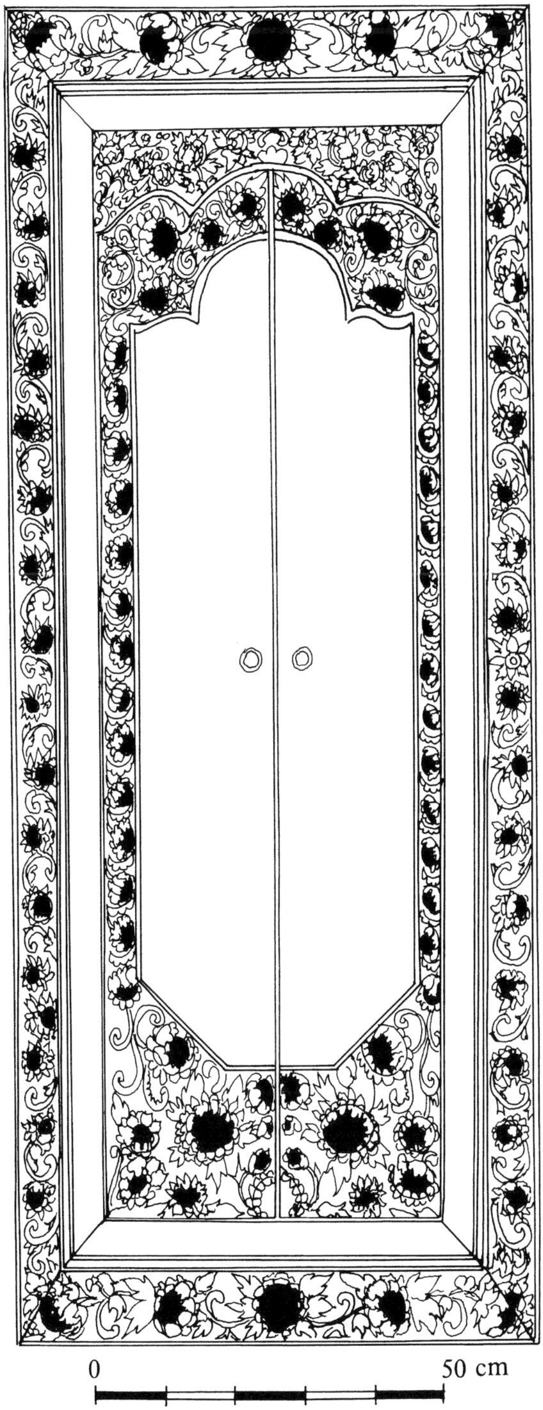

Fig. 52. Bali, door of a house in the village of Ubud (1979).

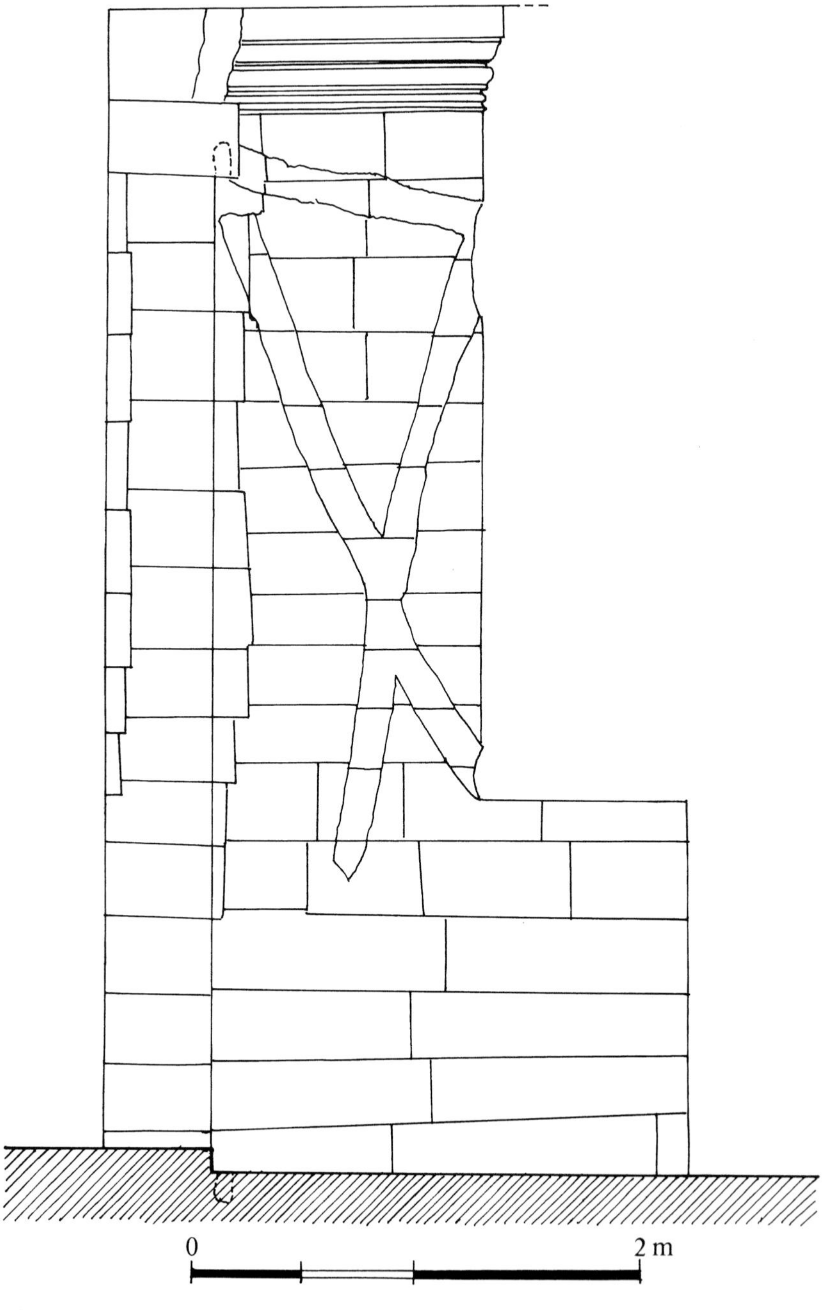

Fig. 53. Angkor, temple of Ta Prohm, jamb of the northern door showing the casement of the repaired leaf.

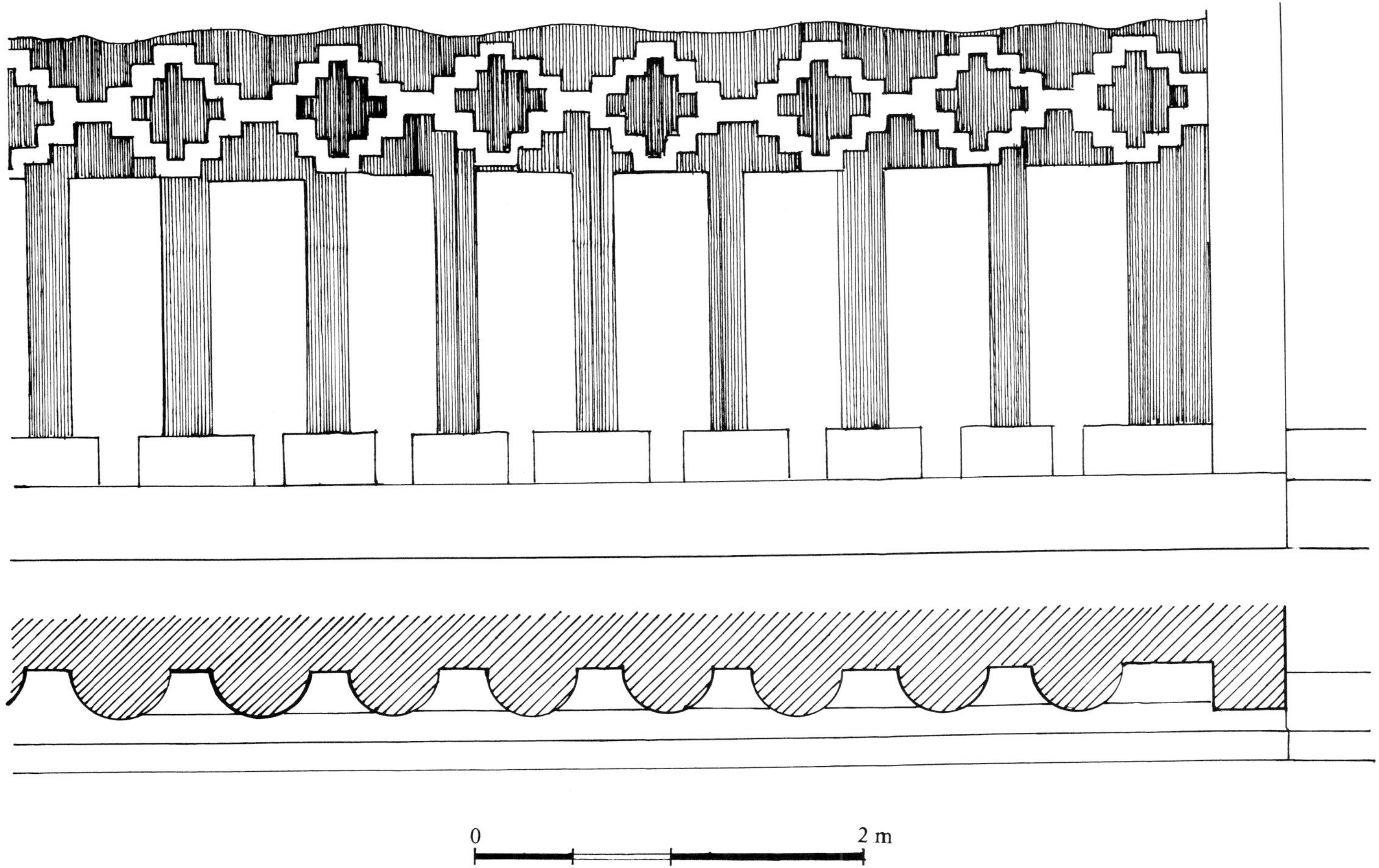

Fig. 54. Mundigak, Afghanistan, plan and elevation of the wall of the palace.

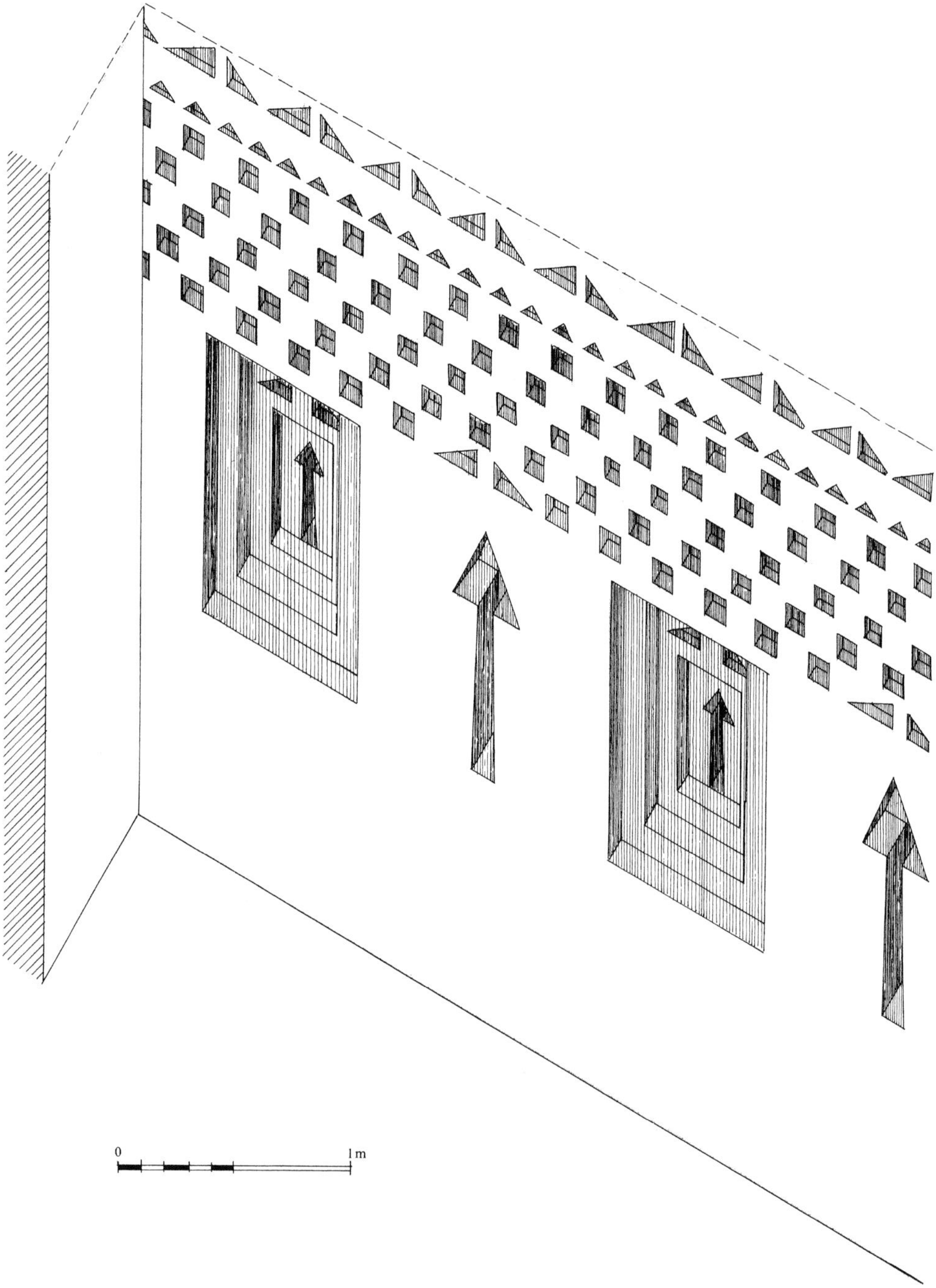

Fig. 55. Surkh Kotal, Afghanistan, elevation of the décor of the interior enclosure wall.

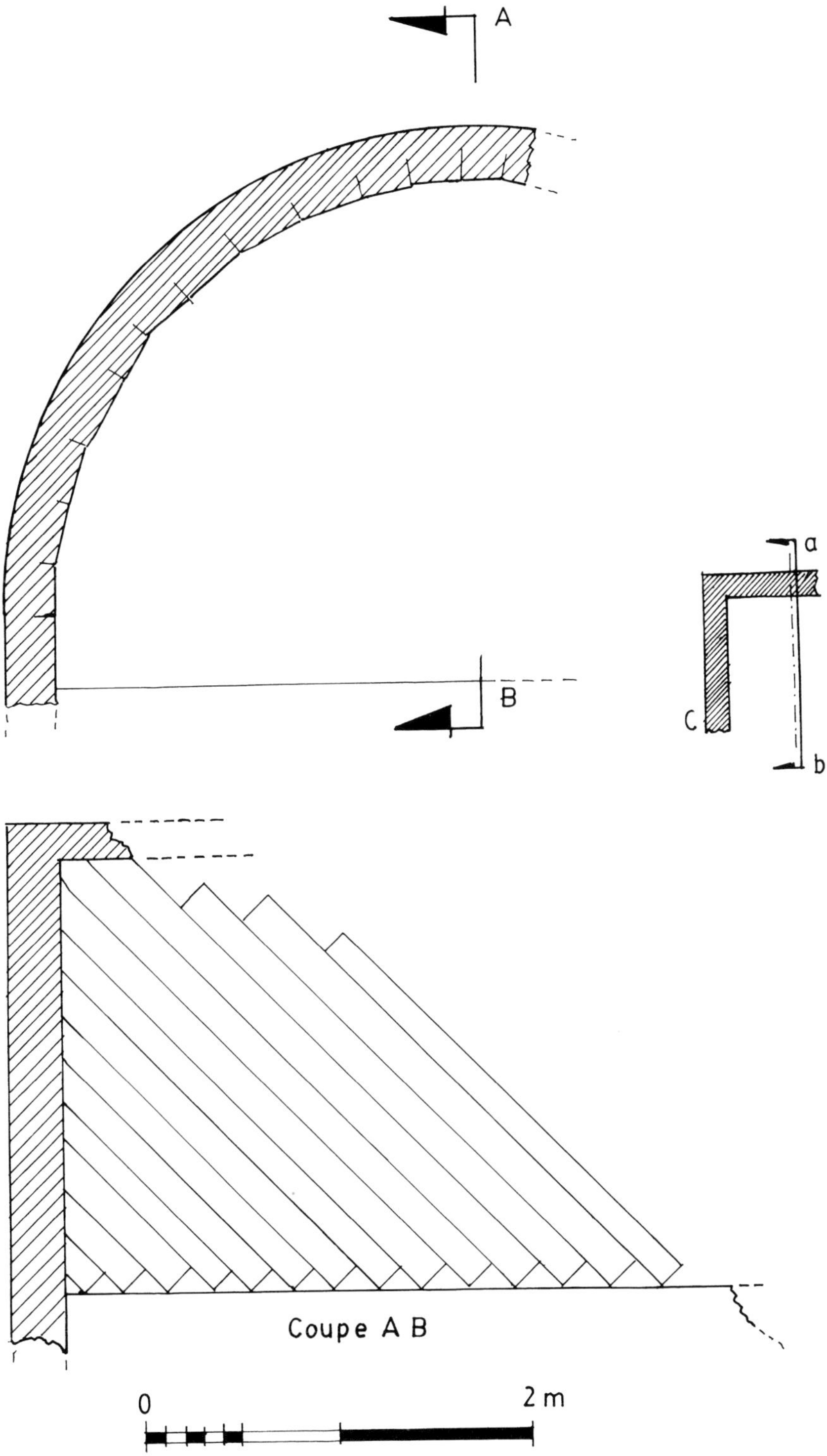

Fig. 56. Barrel vault constructed with unburnt bricks.

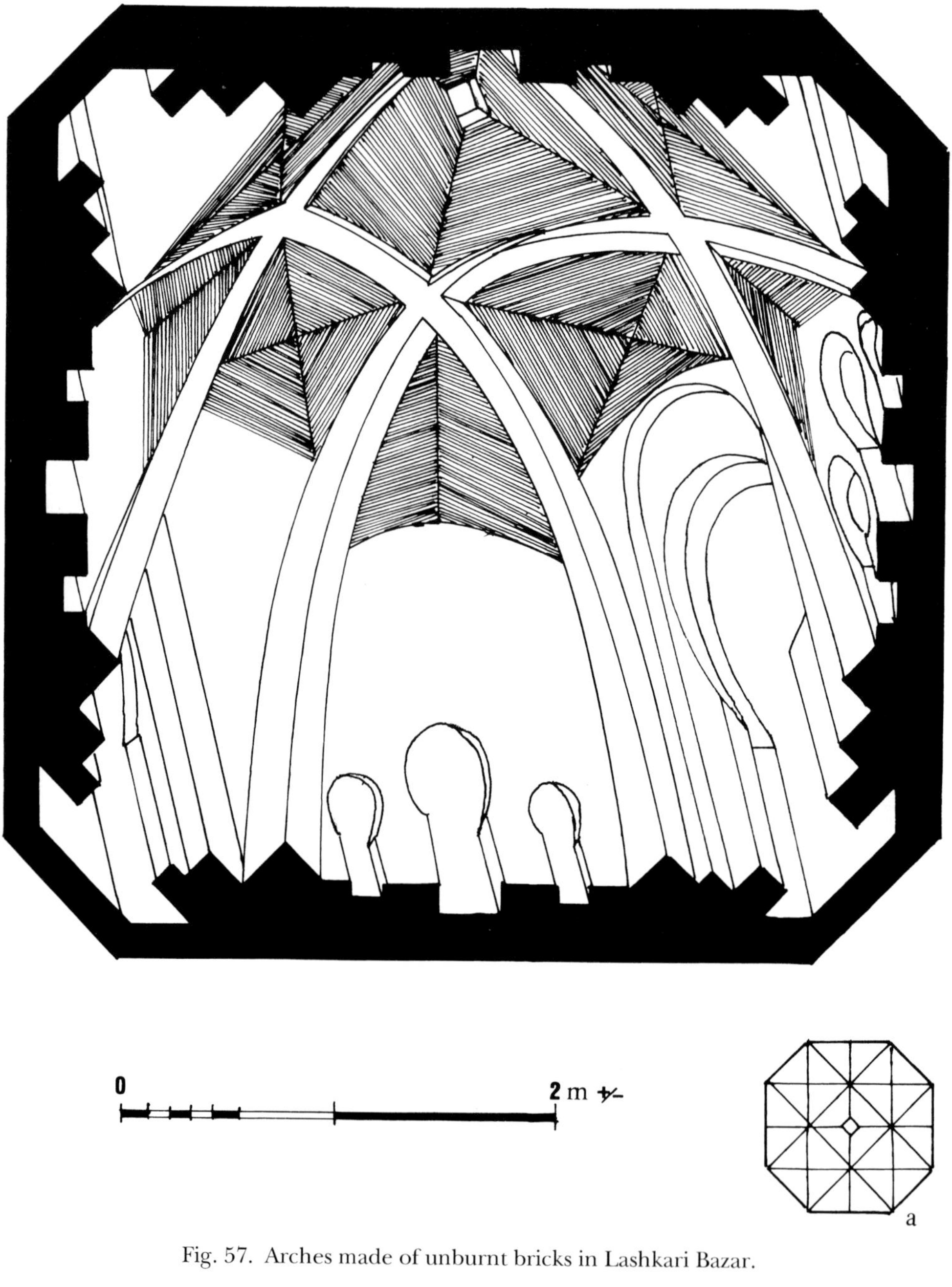

Fig. 57. Arches made of unburnt bricks in Lashkari Bazar.

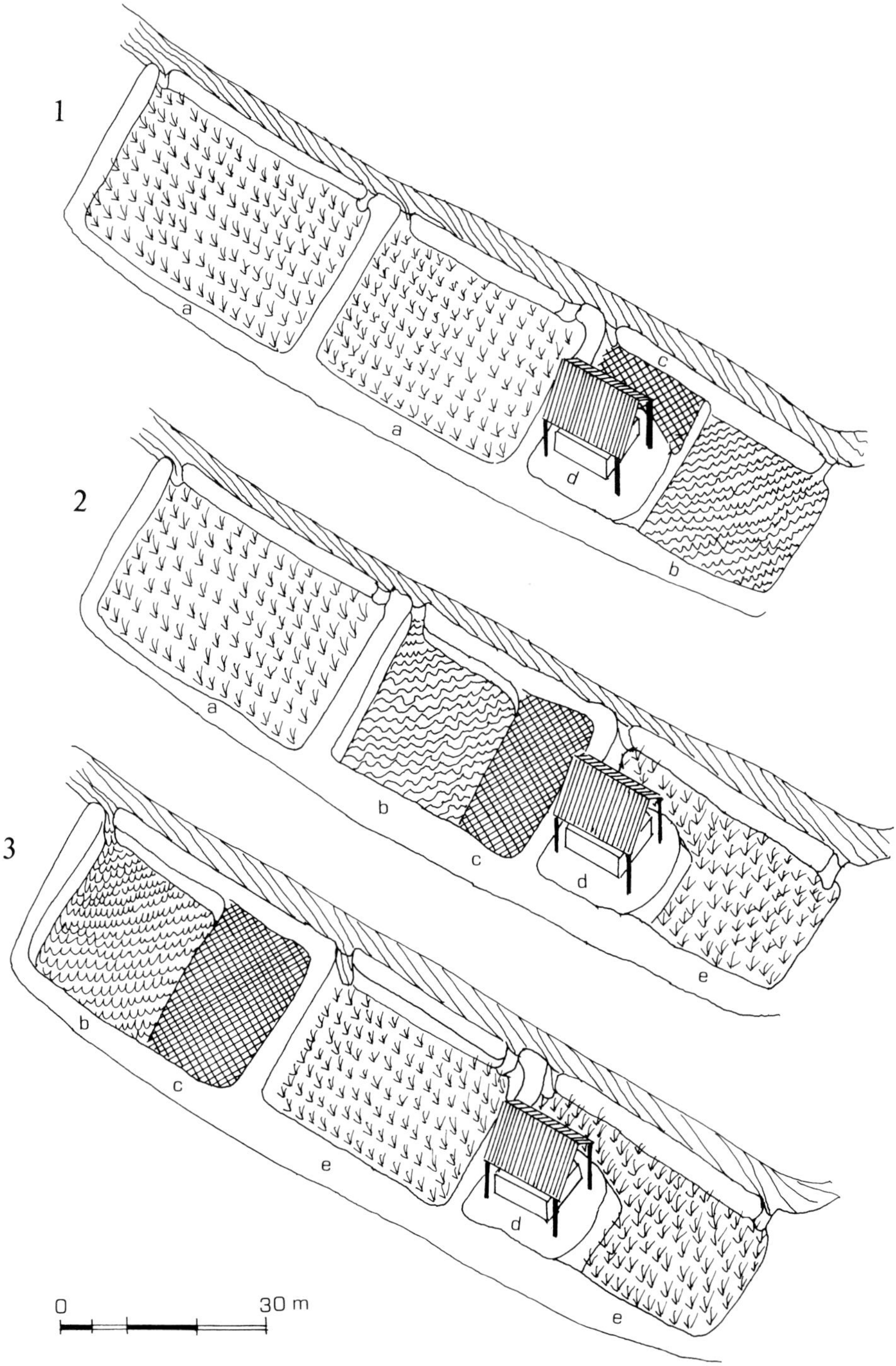

Fig. 58. Creation of bricks in Sawanti (village near Magelang, Java, 1977-1978) at the period of the drop in the rice paddy water levels. Legend: a, rice paddies at normal level before work; b, area where the paste is prepared; c, brick drying area; d, oven; e, rice paddies at normal level after the work.

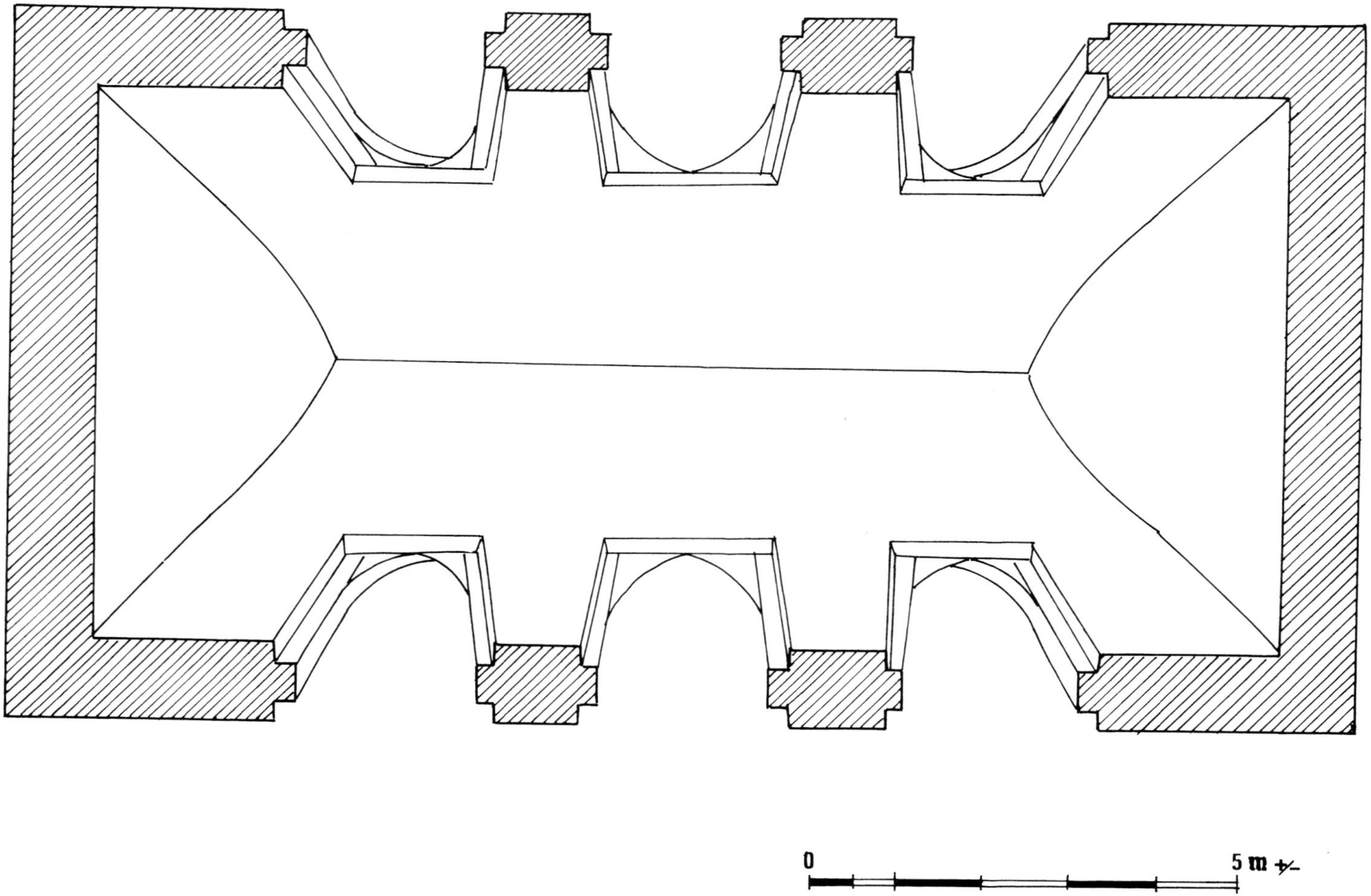

Fig. 59. Perspective of the intrados of the brick vault of a temple of Bengal, “bangla” style.

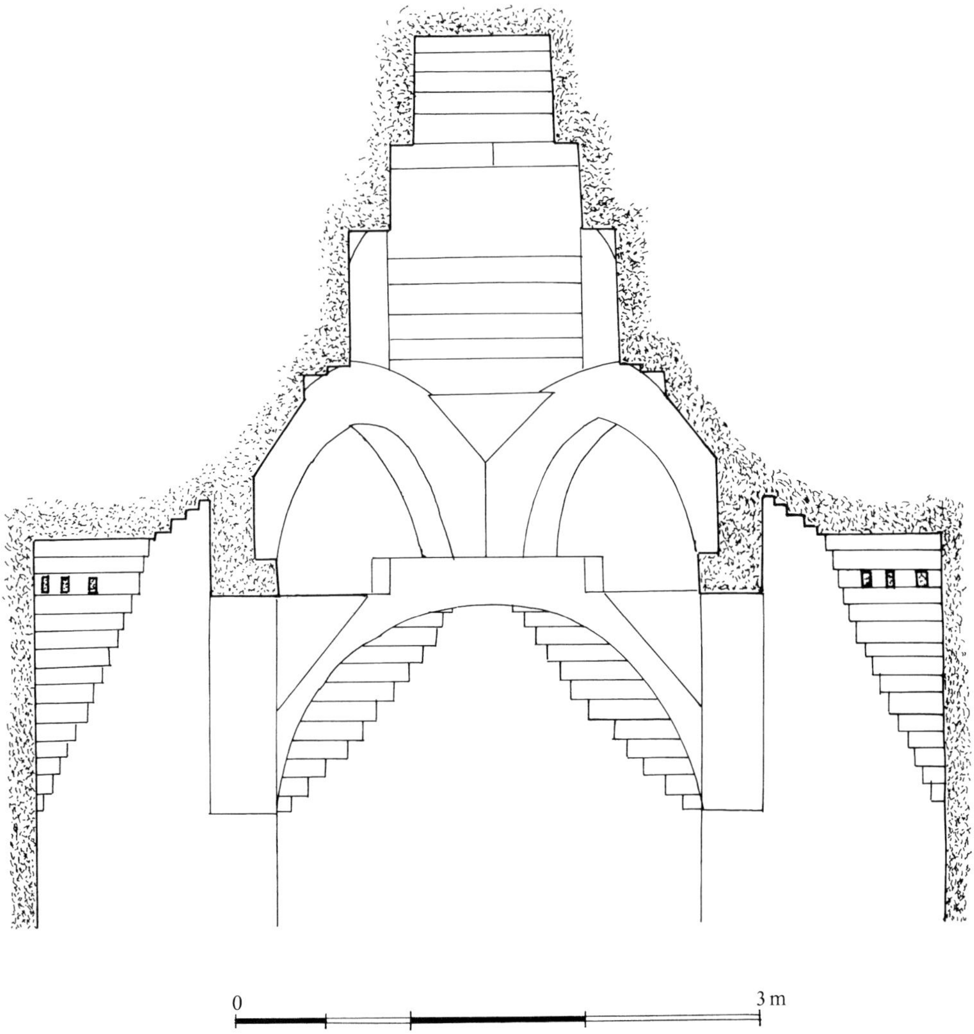

Fig. 60. Darasuram, temple of Amman, cross section of the cella ceiling.

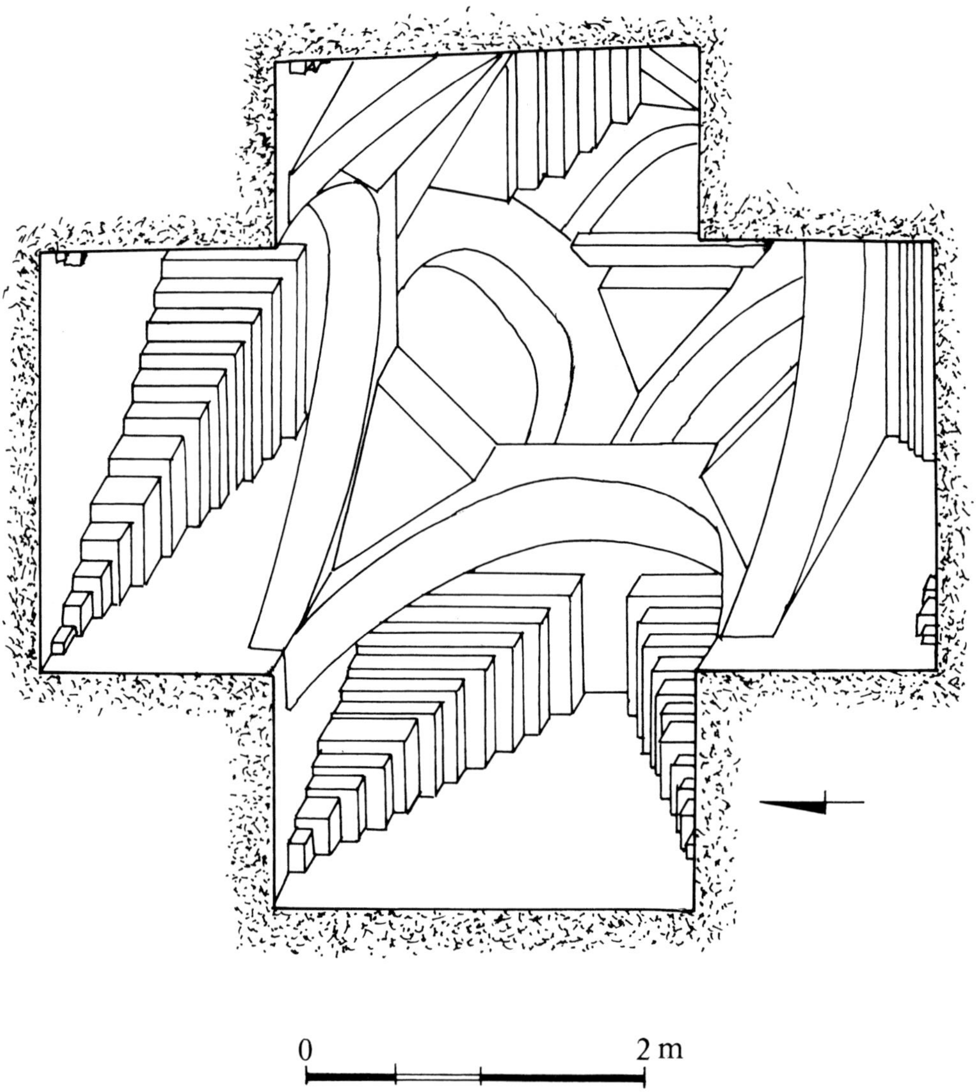

Fig. 61. Darasuram, axonometry of the intrados of the vault of the Amman temple cella.

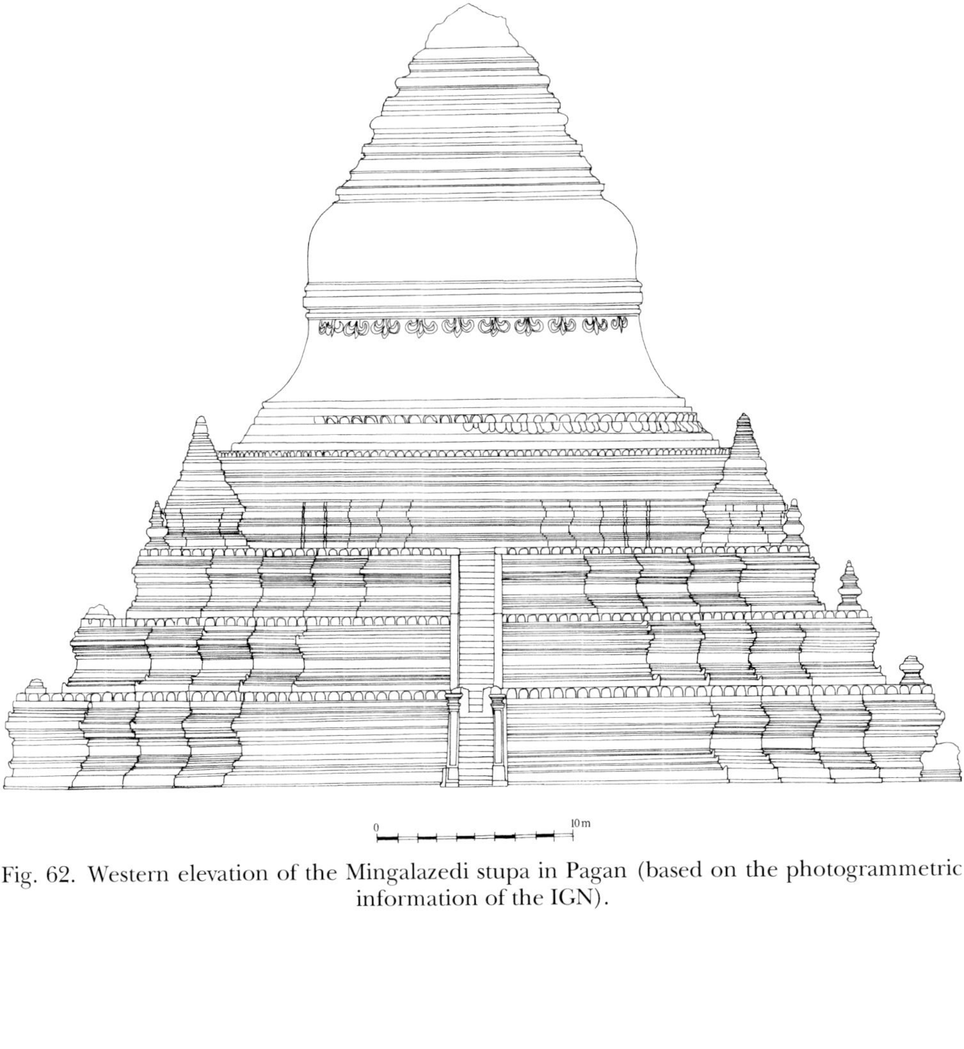

Fig. 62. Western elevation of the Mingalazedi stupa in Pagan (based on the photogrammetric information of the IGN).

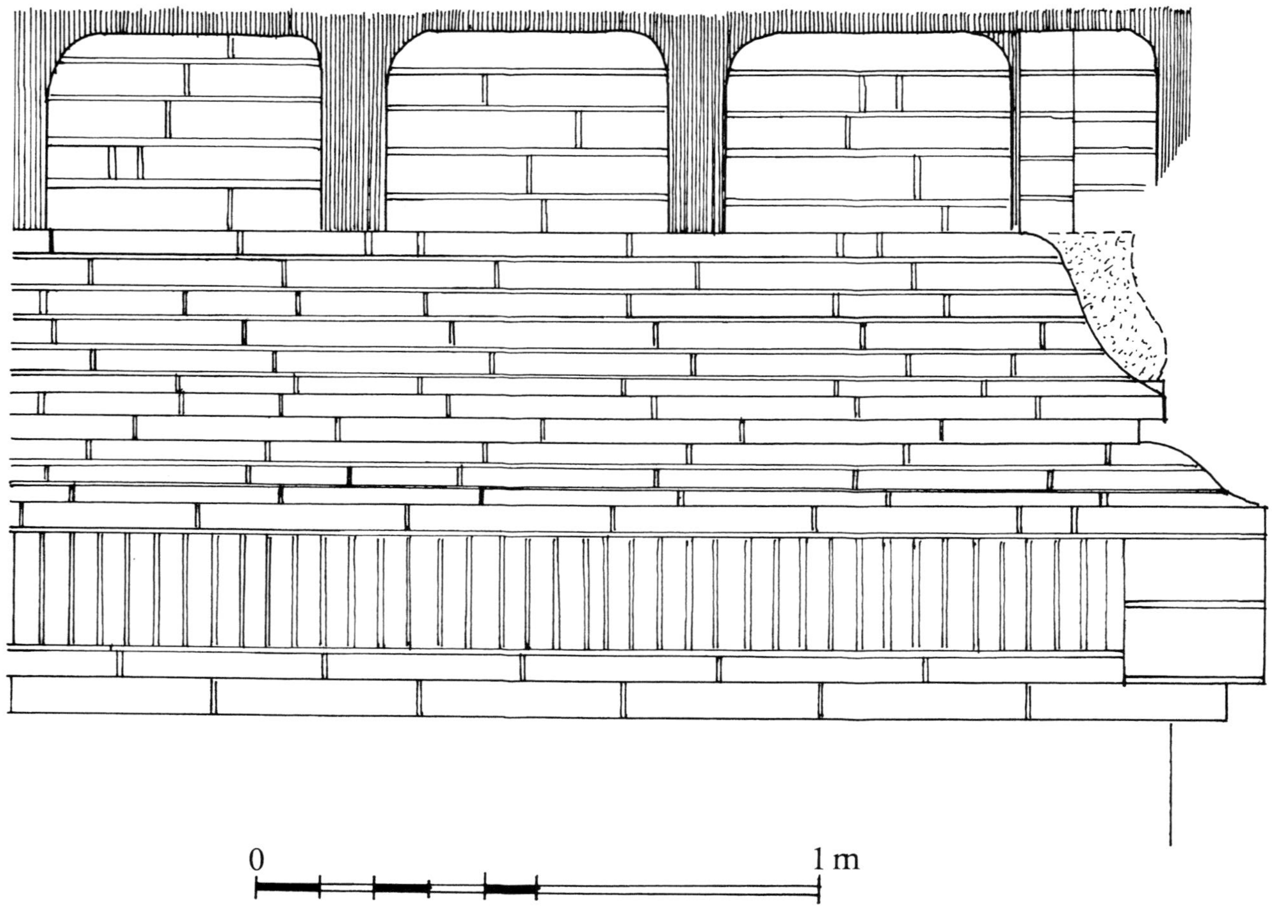

Fig. 63. Mingalazedi, masonry detail of the capping of the first floor, western façade.

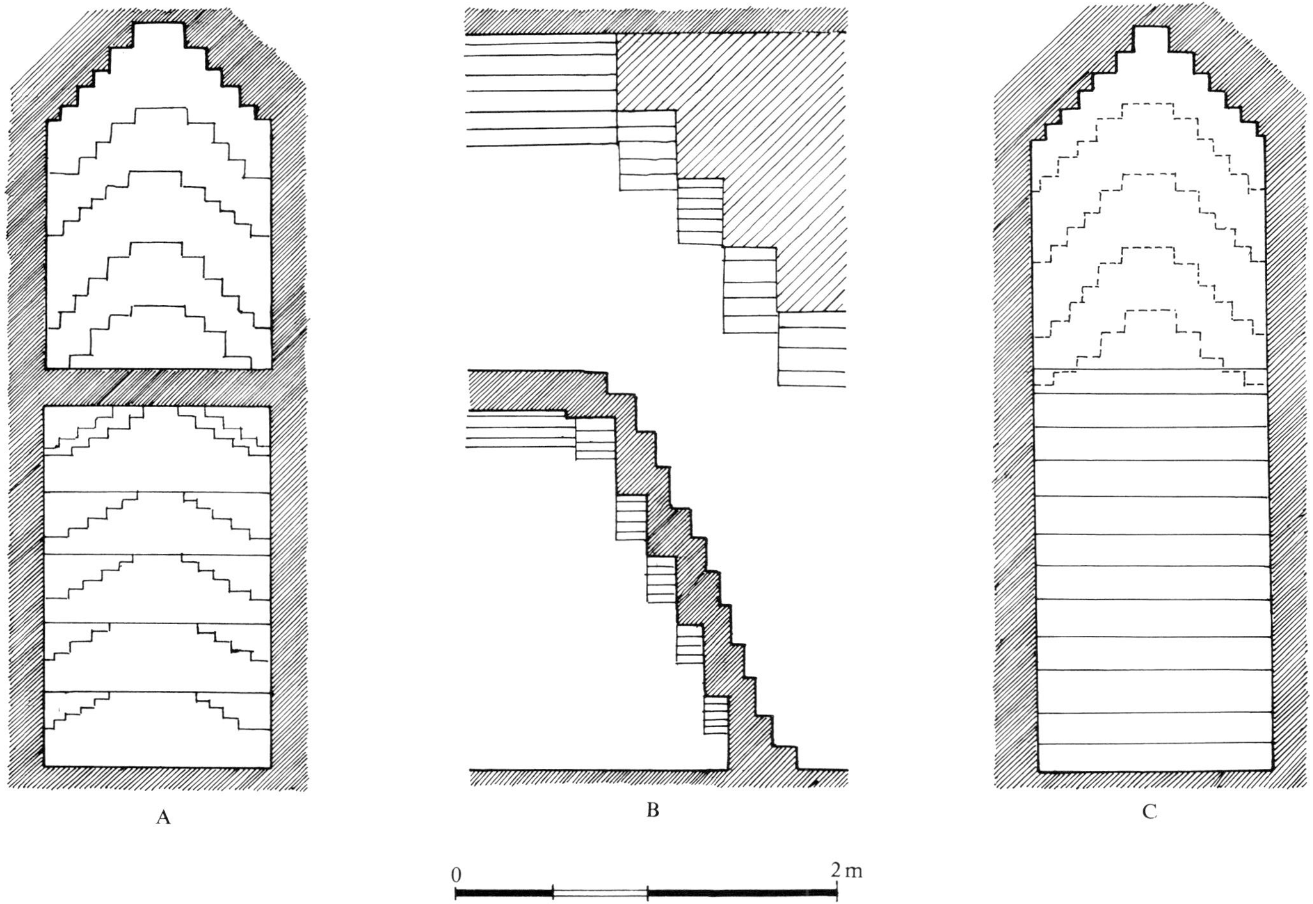

Fig. 64. Pagan, restitution of the string and roofing of a staircase, based on structure No. 2182 from the inventory taken by Pierre Pichard, op. cit., vol. VIII.

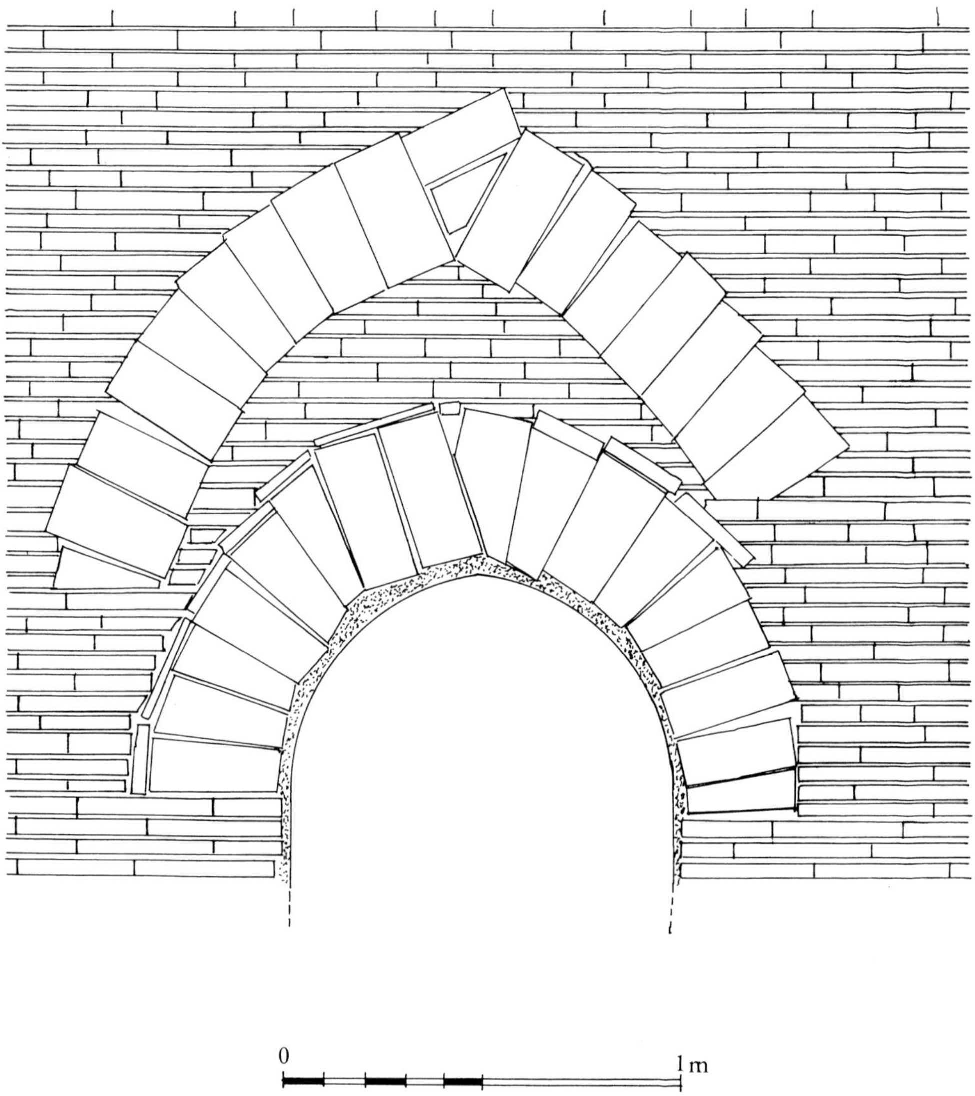

Fig. 65. Brick arch and its discharge, based on sketches made on site and photographs, in particular structure No. 484, Mingala-the-hpaya, Burma, of the inventory taken by P. Pichard, op. cit., vol. II.

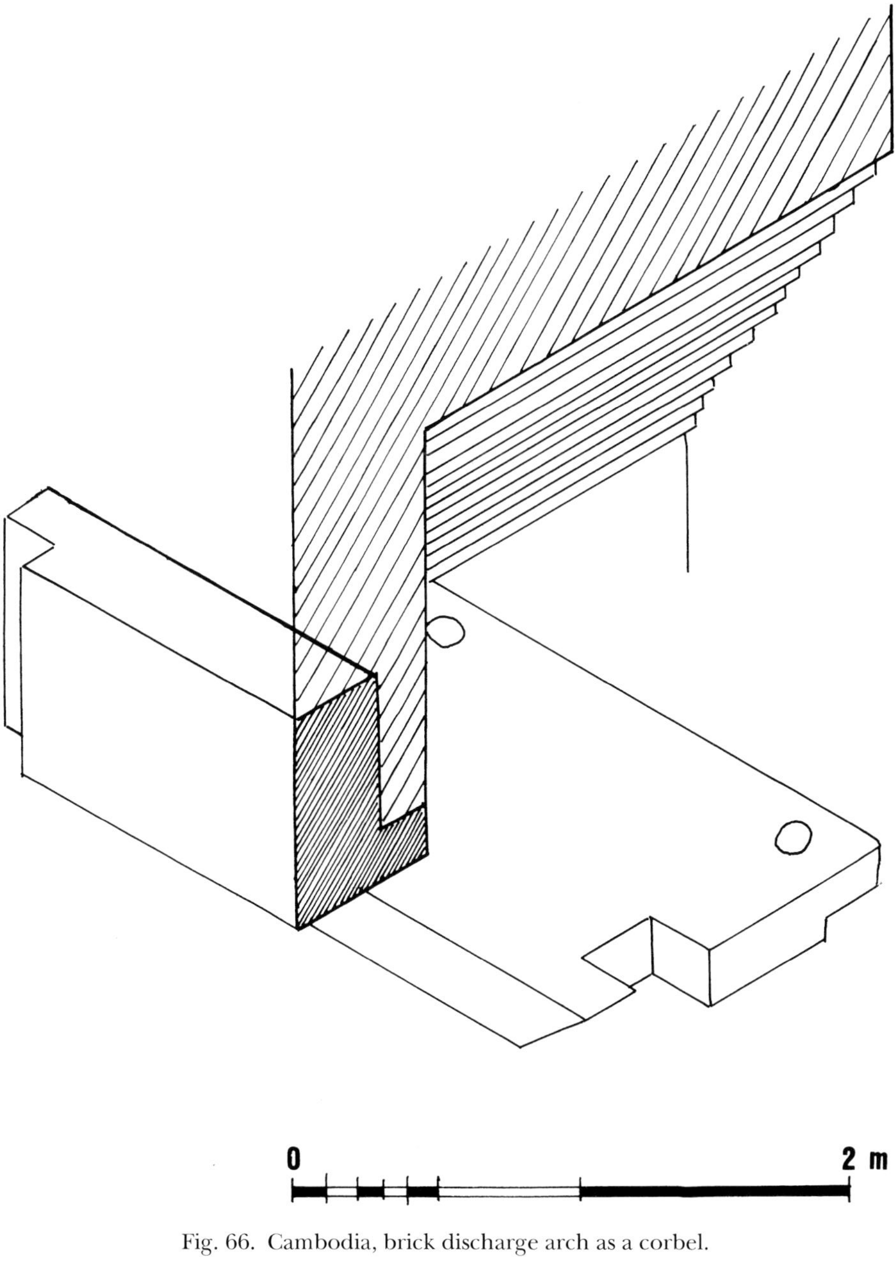

Fig. 66. Cambodia, brick discharge arch as a corbel.

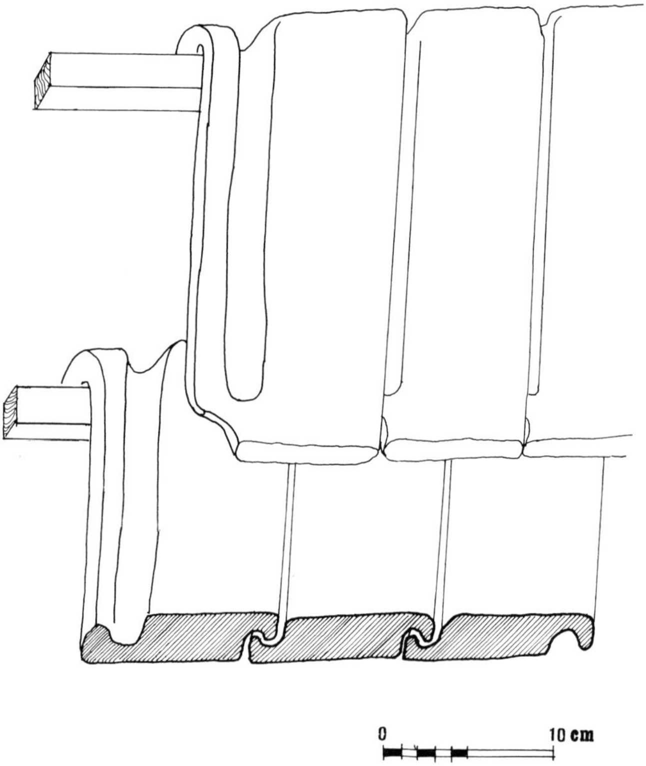

Fig. 67. Pagan, side-fitting tiles.

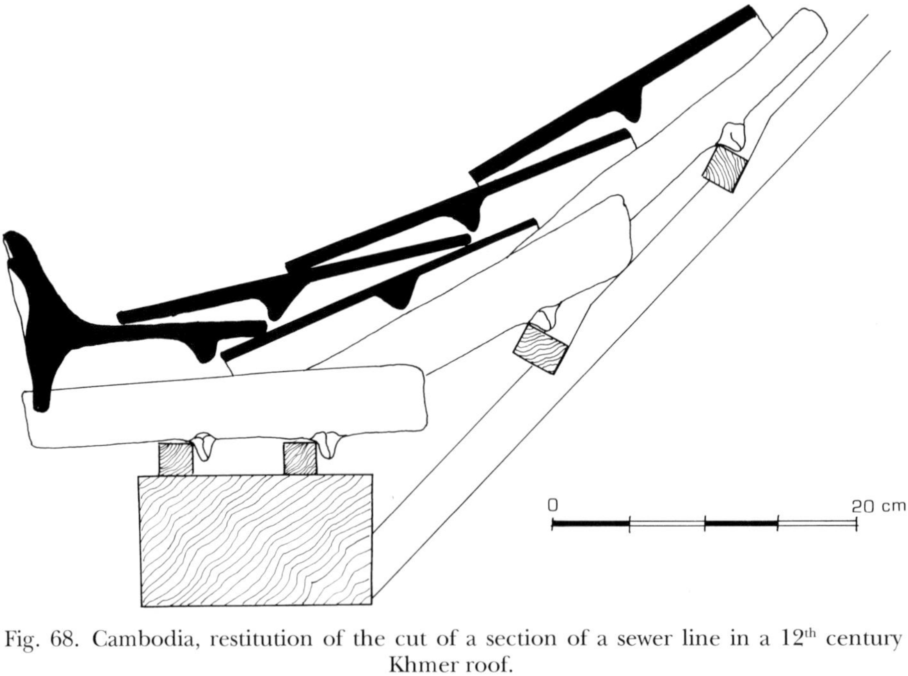

Fig. 68. Cambodia, restitution of the cut of a section of a sewer line in a 12th century Khmer roof.

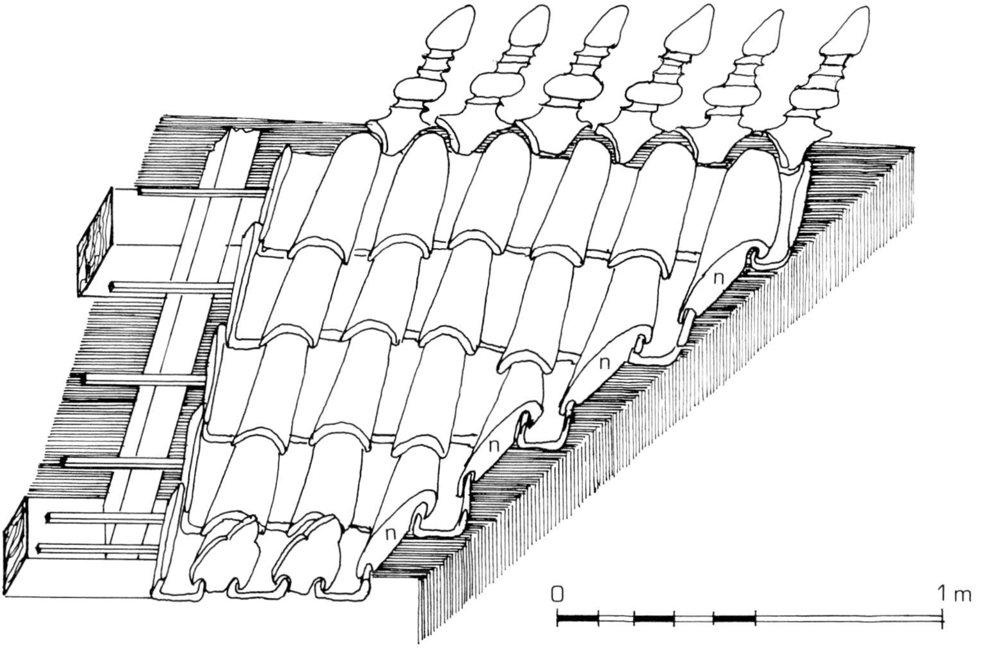

Fig. 69. Cambodia, implementation of valley tiles, marked by the letter "n".

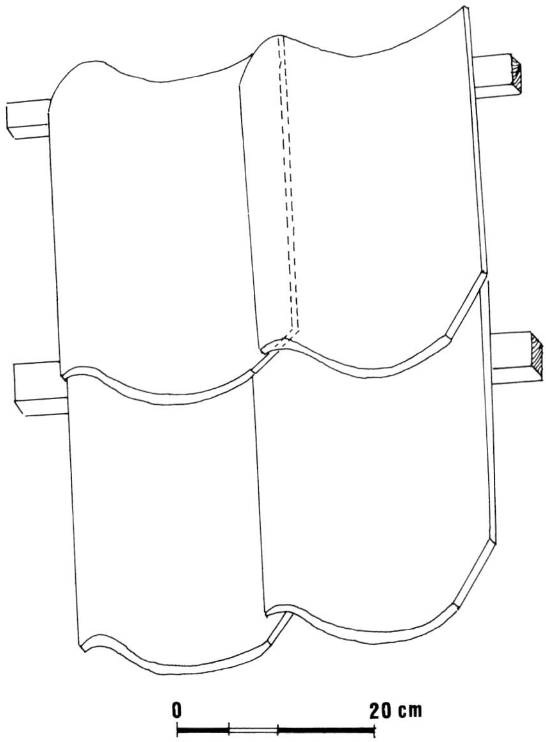

Fig. 70. Java, implementation of double-curved tiles.

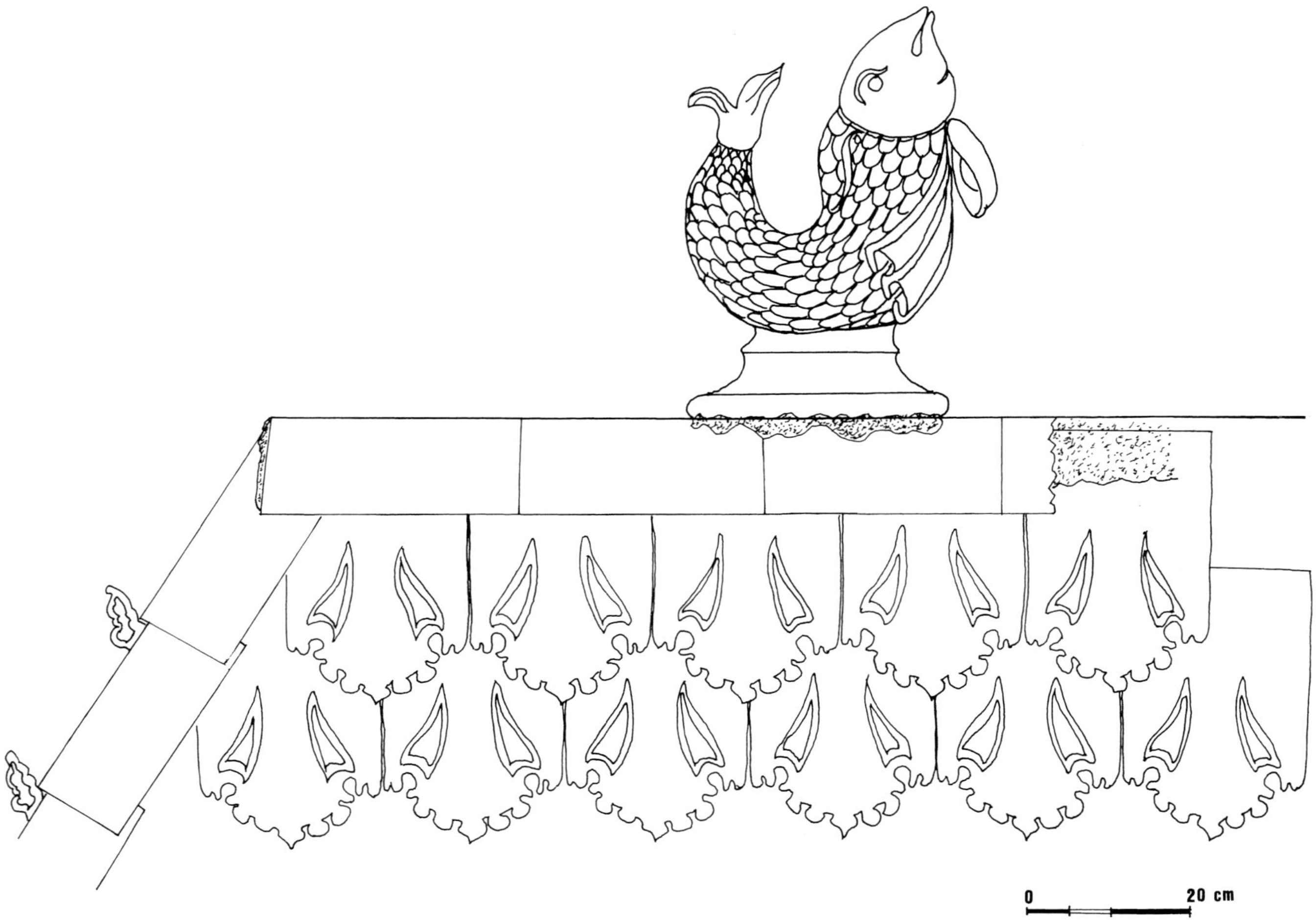

Fig. 71. Java, restitution of roofing from the Mojopahit civilisation, 14th century.

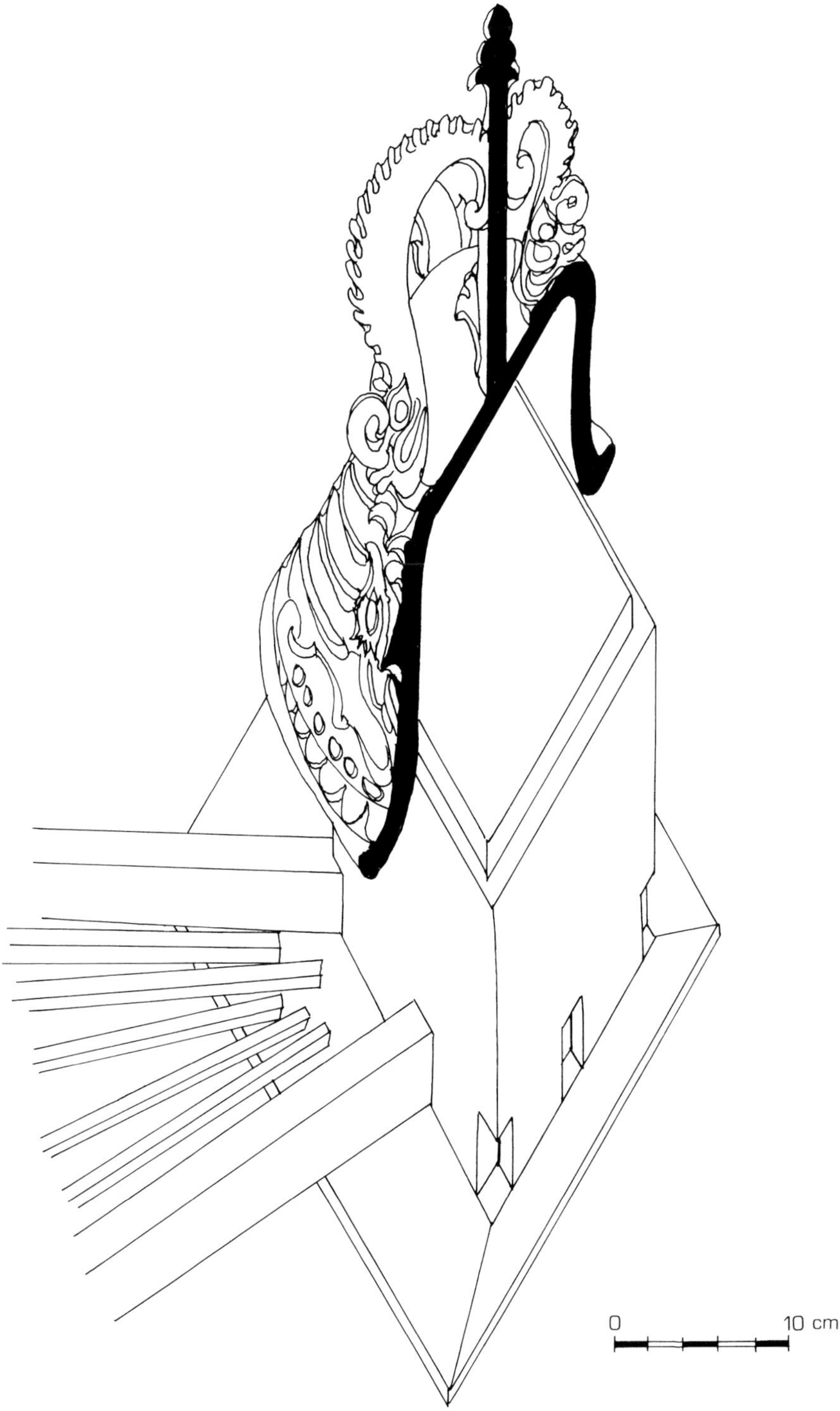

Fig. 72. Bali, implementation of a ridge finial over a radiating frame.

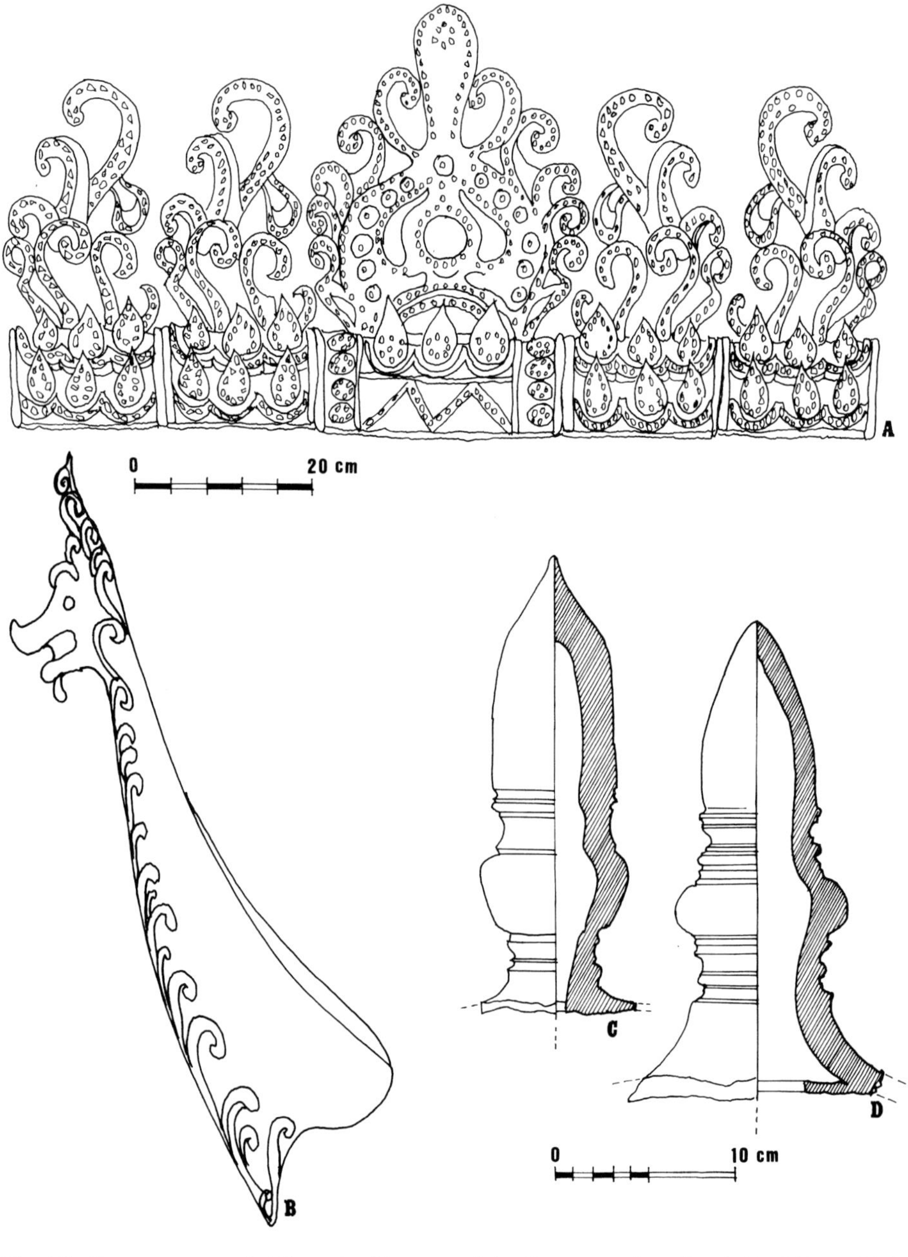

Fig. 73. Terra cotta roof siding: A, ridgecap of the Jepara region (Java). B, Chiang Mai, trim of the roof angle. C and D, Khmer ridge finials.

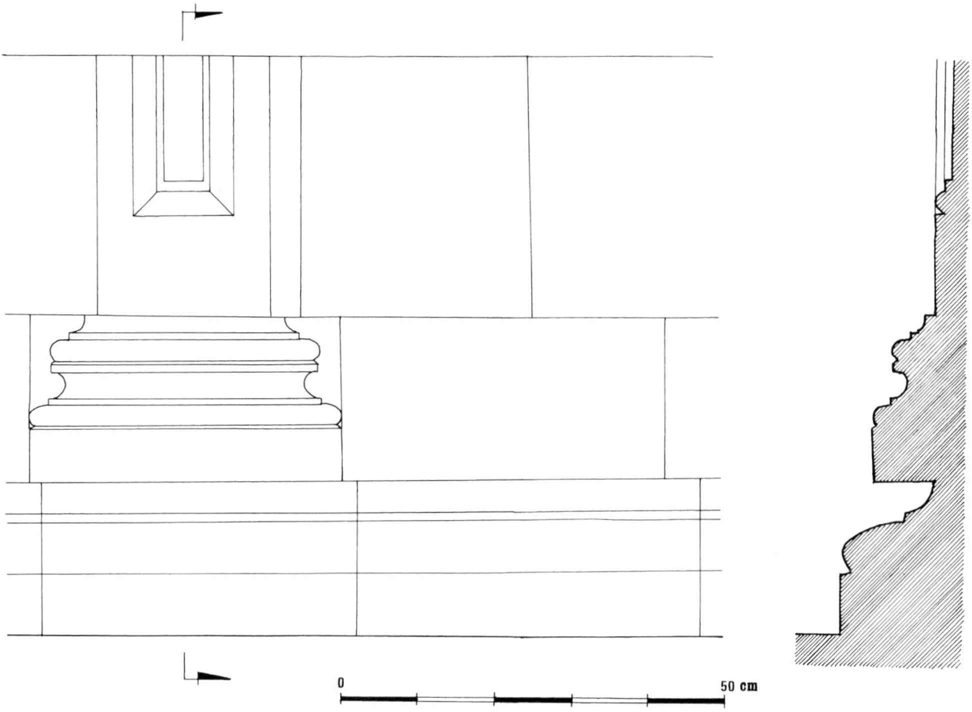

Fig. 74. Surkh Kotal, white lime siding of the foundation of the principal temple.

Fig. 75. Base and cornice of an ionic column of the Jandial temple, near Taxila, Pakistan.

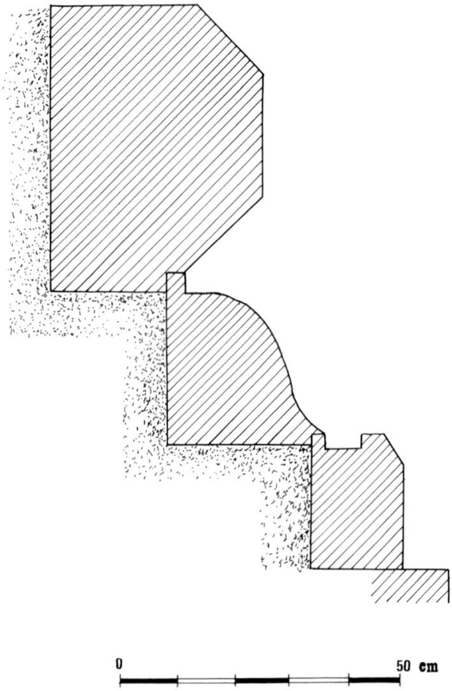

Fig. 76. Stereotomy of the foundation of the Panamalai temple, India.

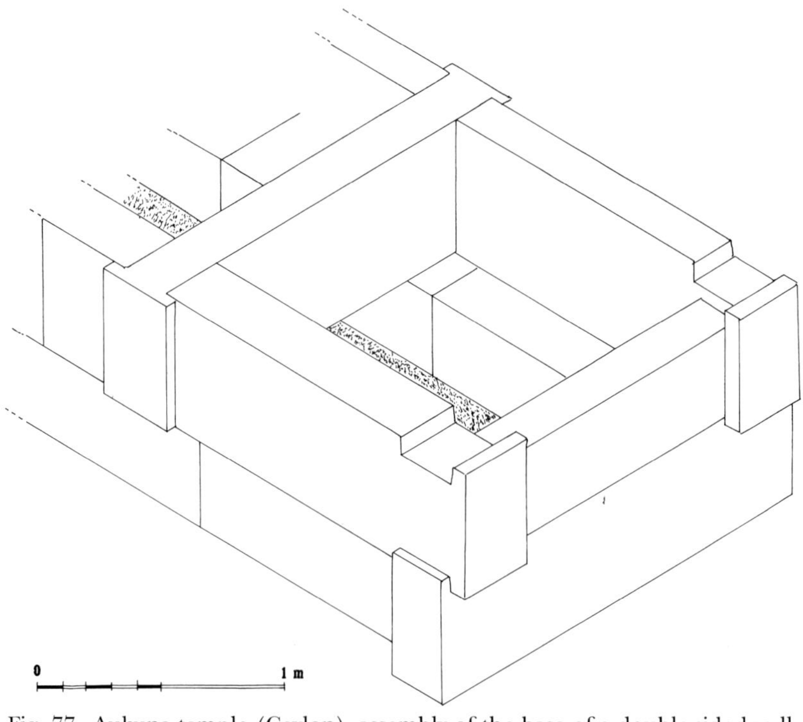

Fig. 77. Aukuna temple (Ceylon), assembly of the base of a double-sided wall, near the principal sanctuary door.

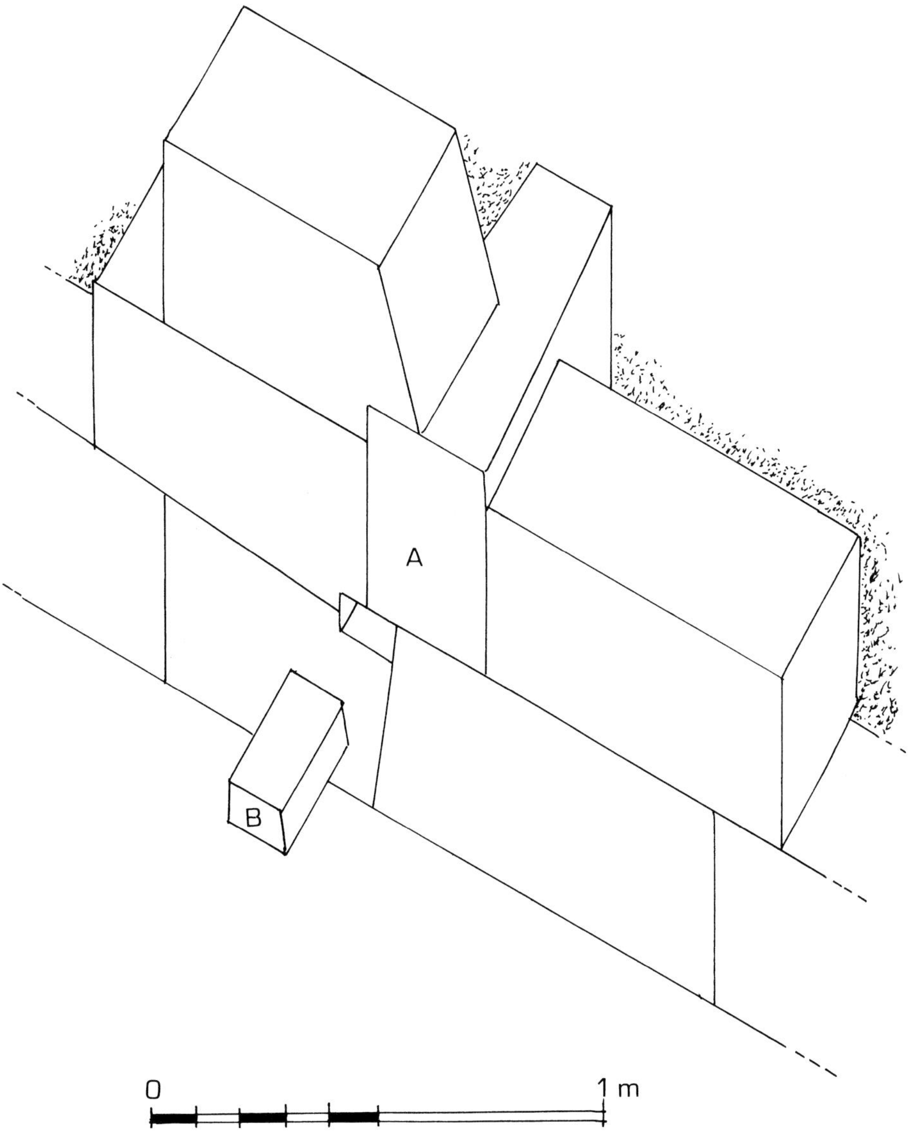

Fig. 78. Stereotomy of the base of the foundation of a building in Mihintale (Ceylon). A, striking wedge in the horizontal plane; B, wedge put into place after the course was adjusted.

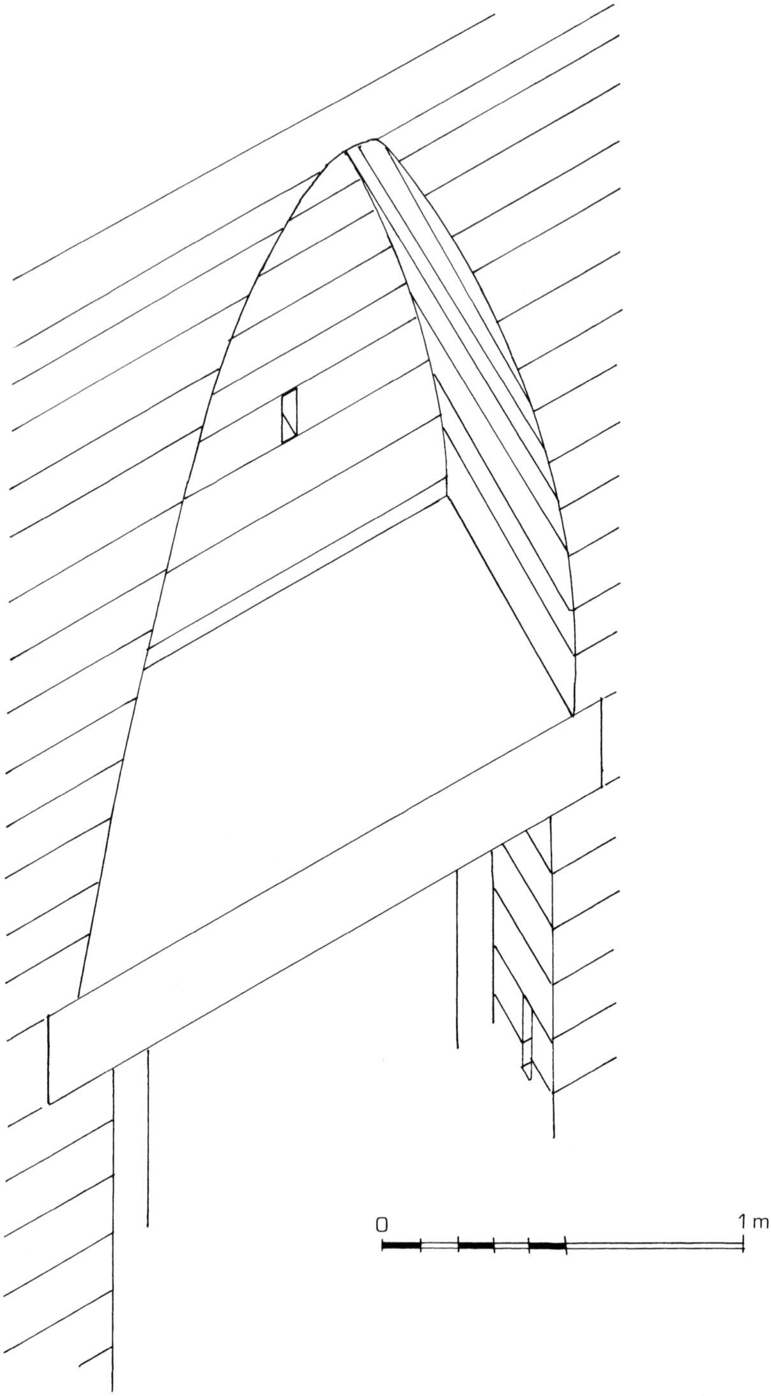

Fig. 79. Discharge arch above the lintel of a door at Candi Bhima in Dieng, Java.

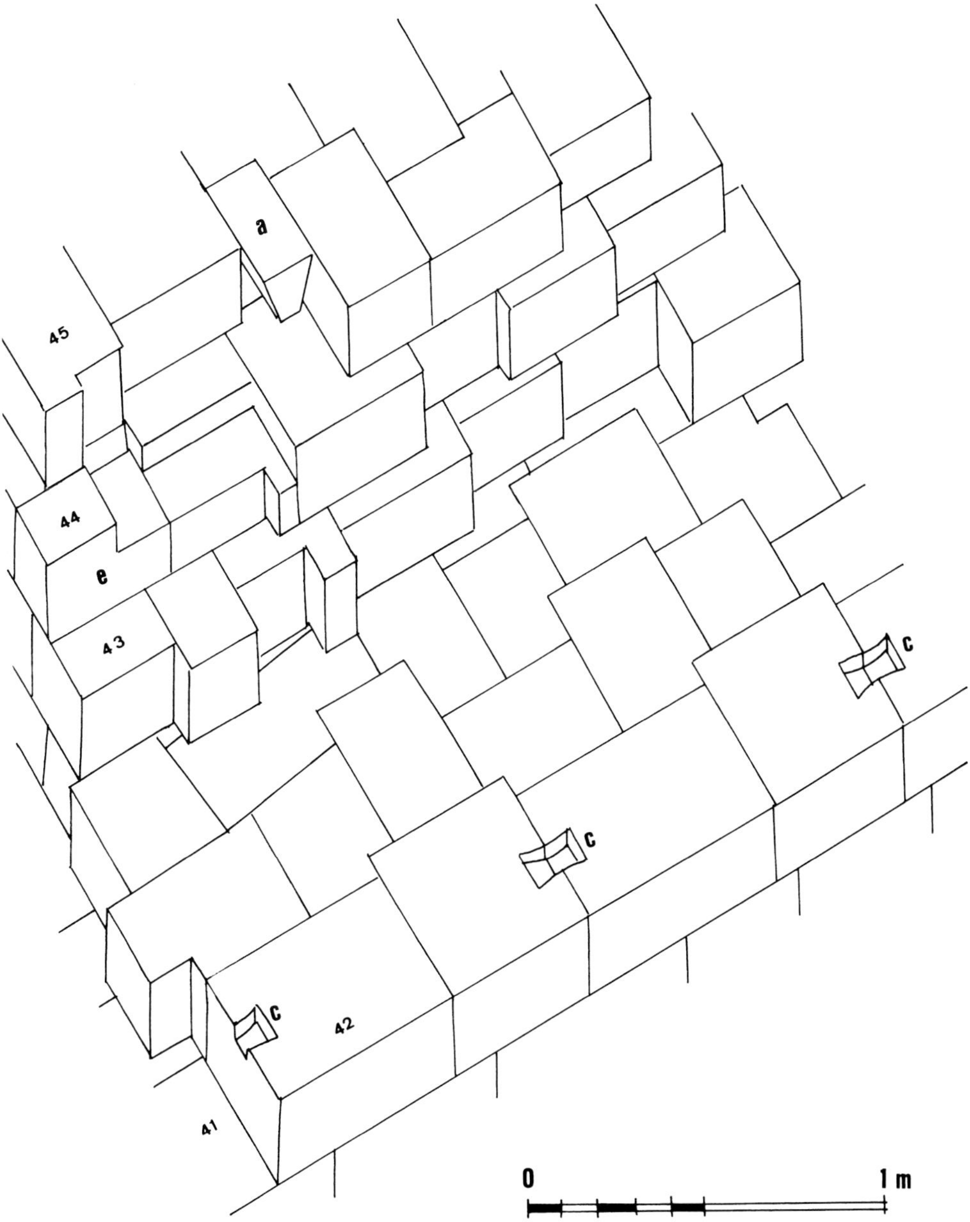

Fig. 80. Borobudur, courses 41 and 45 of the internal base of the first gallery, implementation of double dovetail spikes, C, of the representation (based on pl. LII of *L'Histoire architecturale du Borobudur*, op. cit.).

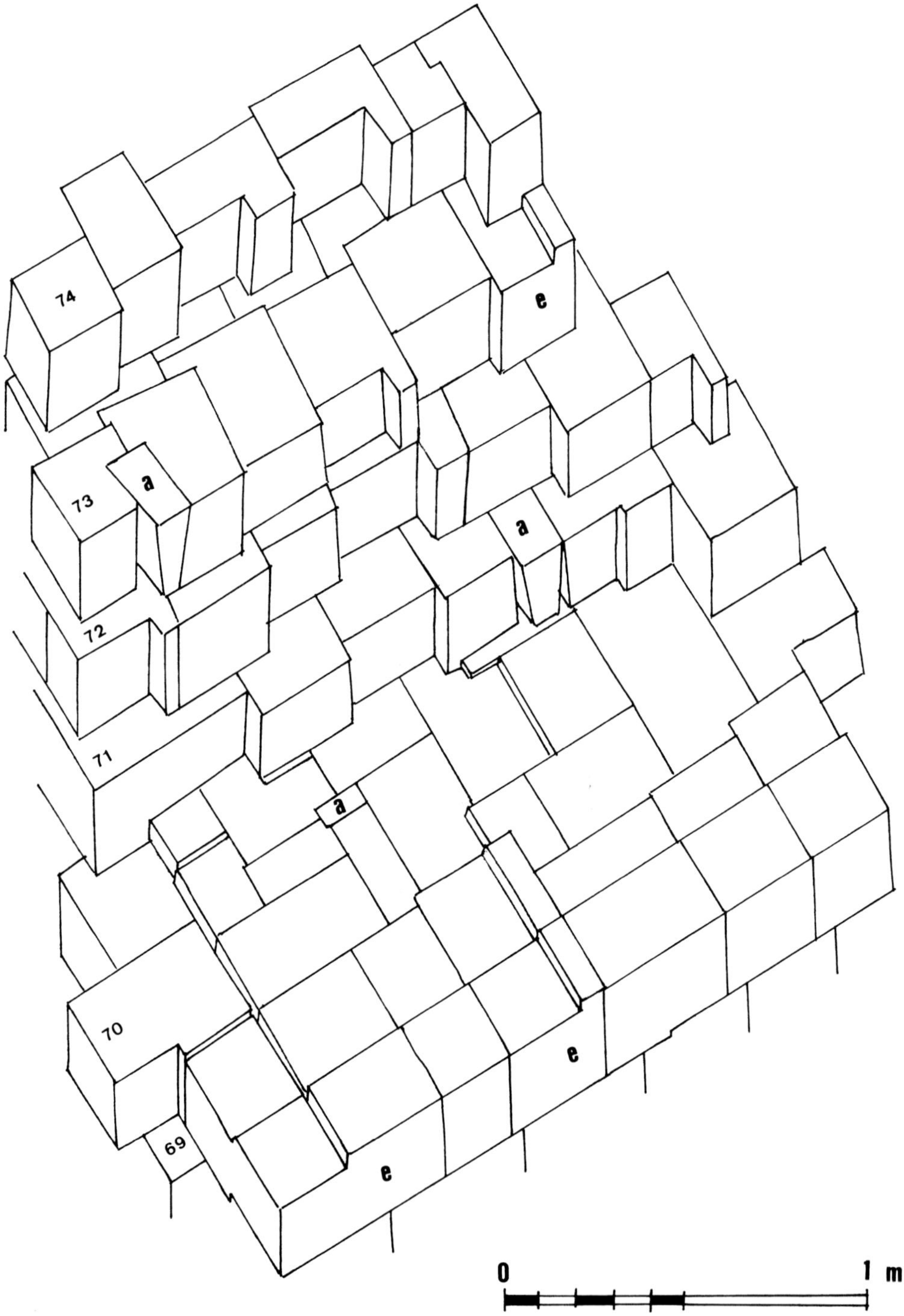

Fig. 81. Borobudur, stereotomy of the internal base on the level of the third gallery of the 69th to the 74th course, with striking wedge, and angle bar in the vertical plane (based on pl. LIII of *L'Histoire architecturale du Borobudur*, op. cit.).

Fig. 82. Borobudur, passage from the first to the second gallery, crossette keystones.

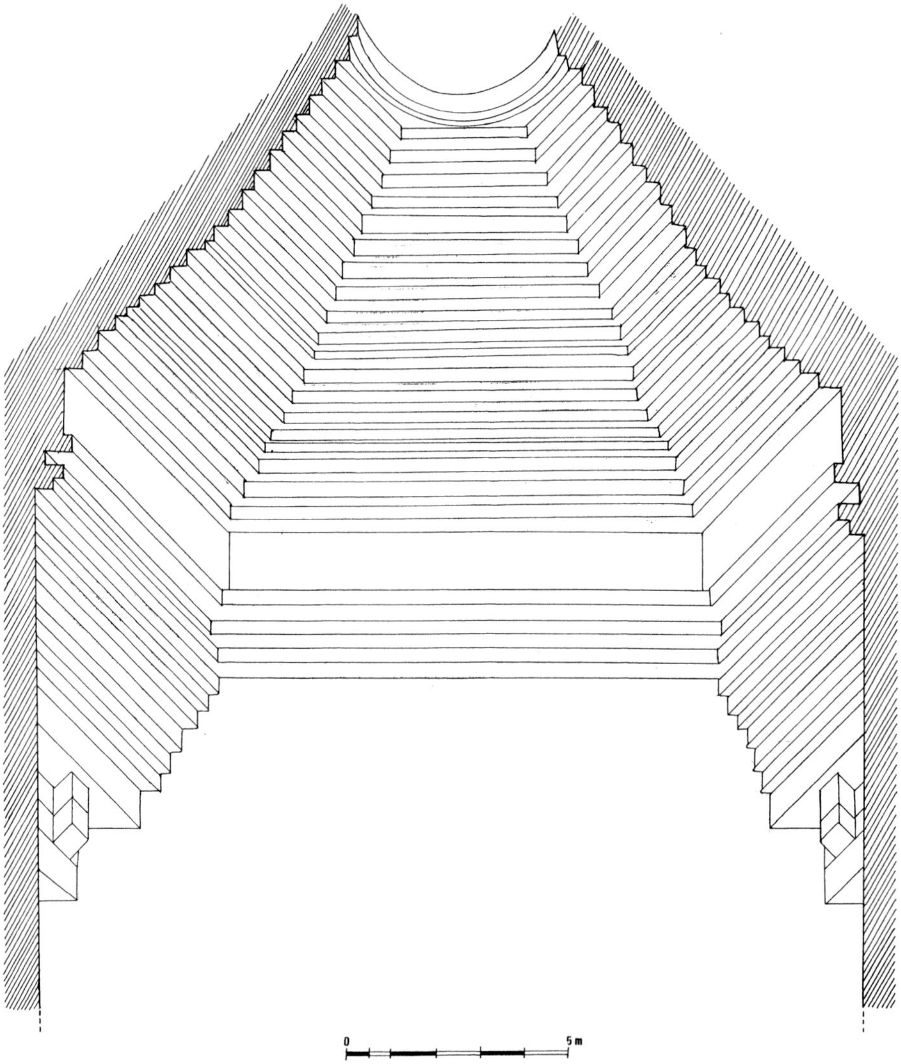

Fig. 83. Candi Kalasan, Java, passage from the square plane to the octagonal plane of the cella (corbel trumpet arches).

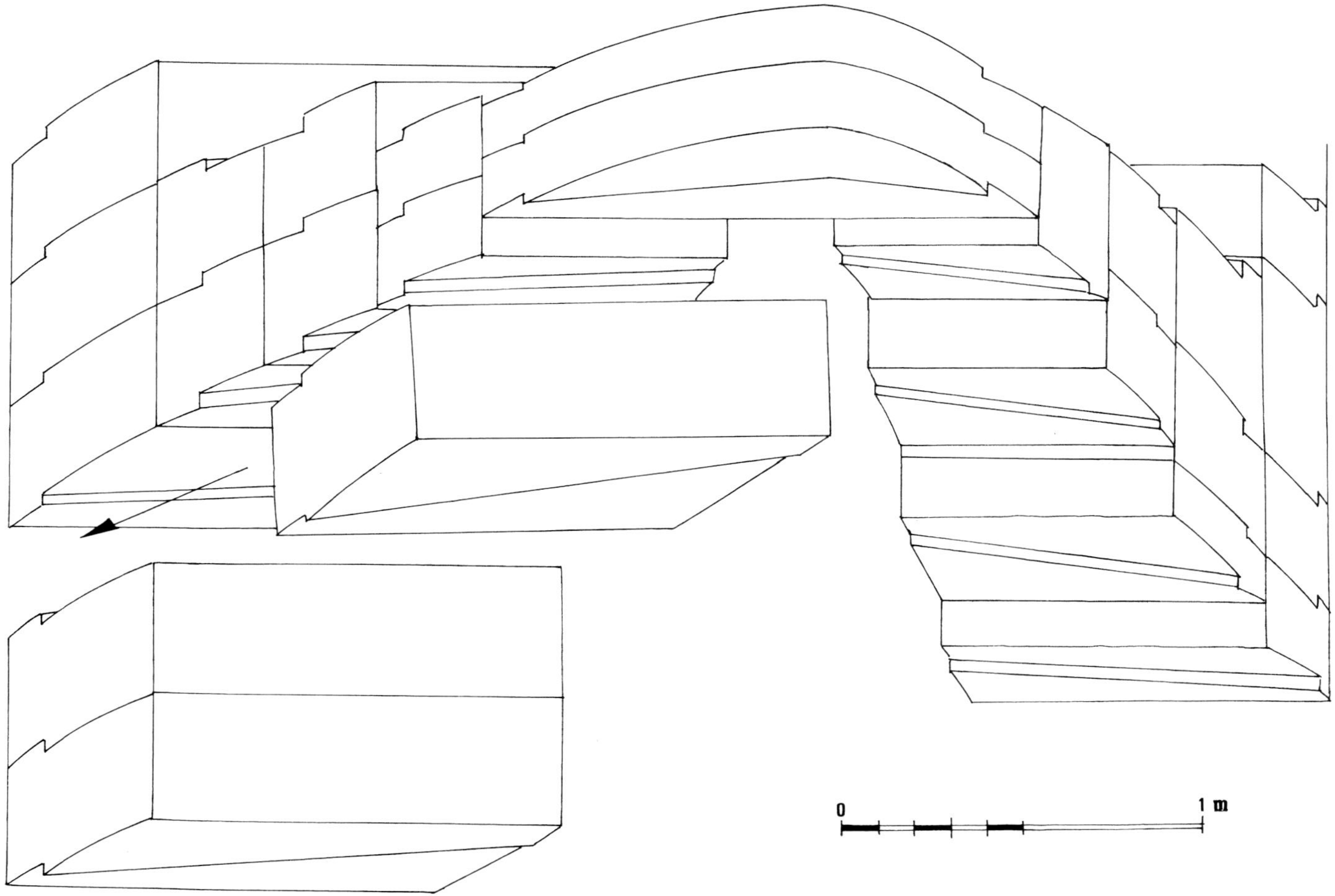

Fig. 84. Angkor Wat, stereotomy of the corbel covering the gallery of the first floor and implementation of a keystone.

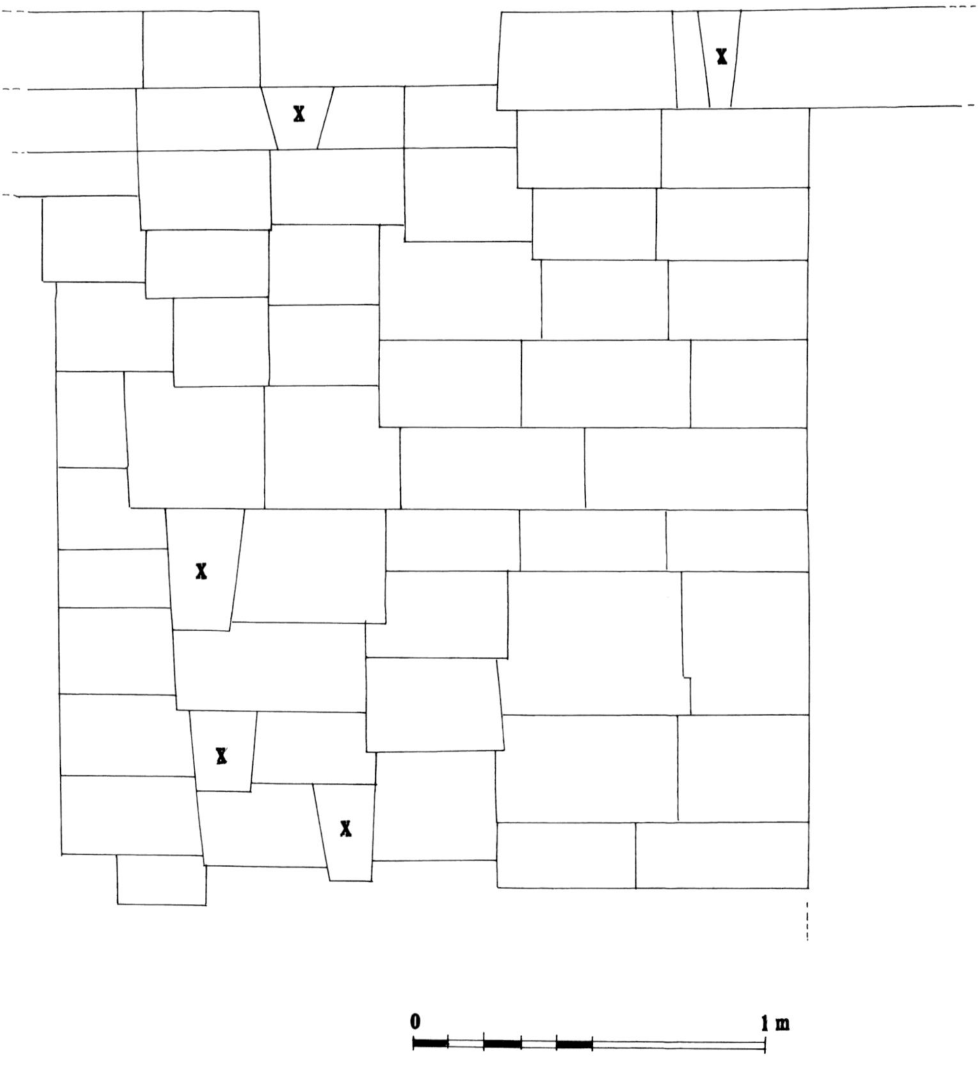

Fig. 85. Bayon, eastern wall of the first gallery, south side, stereotomy detail. The stones marked "X" are striking wedges.

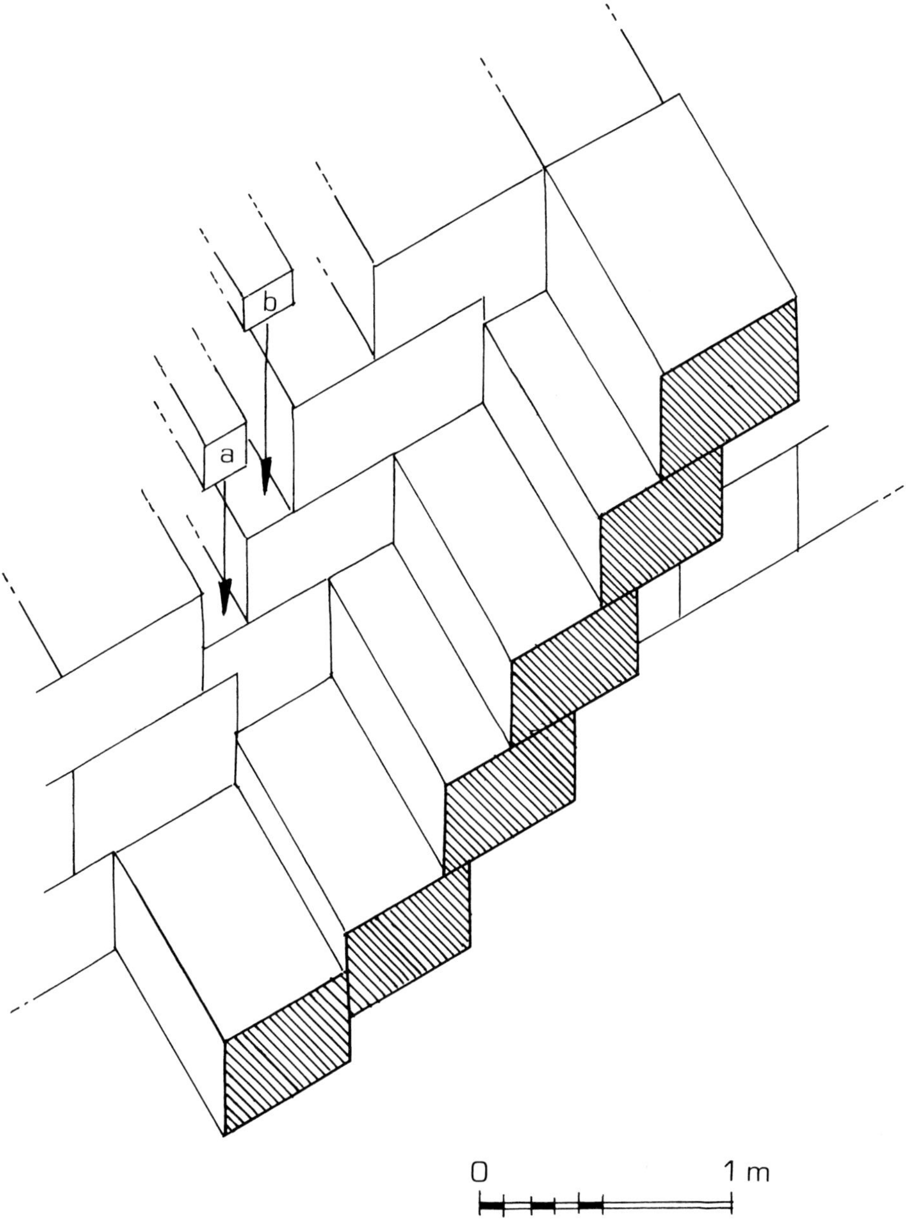

Fig. 86. Darasuram, stereotomy of the northern staircase of the entry pavilion II/east, stones a and b are wedged.

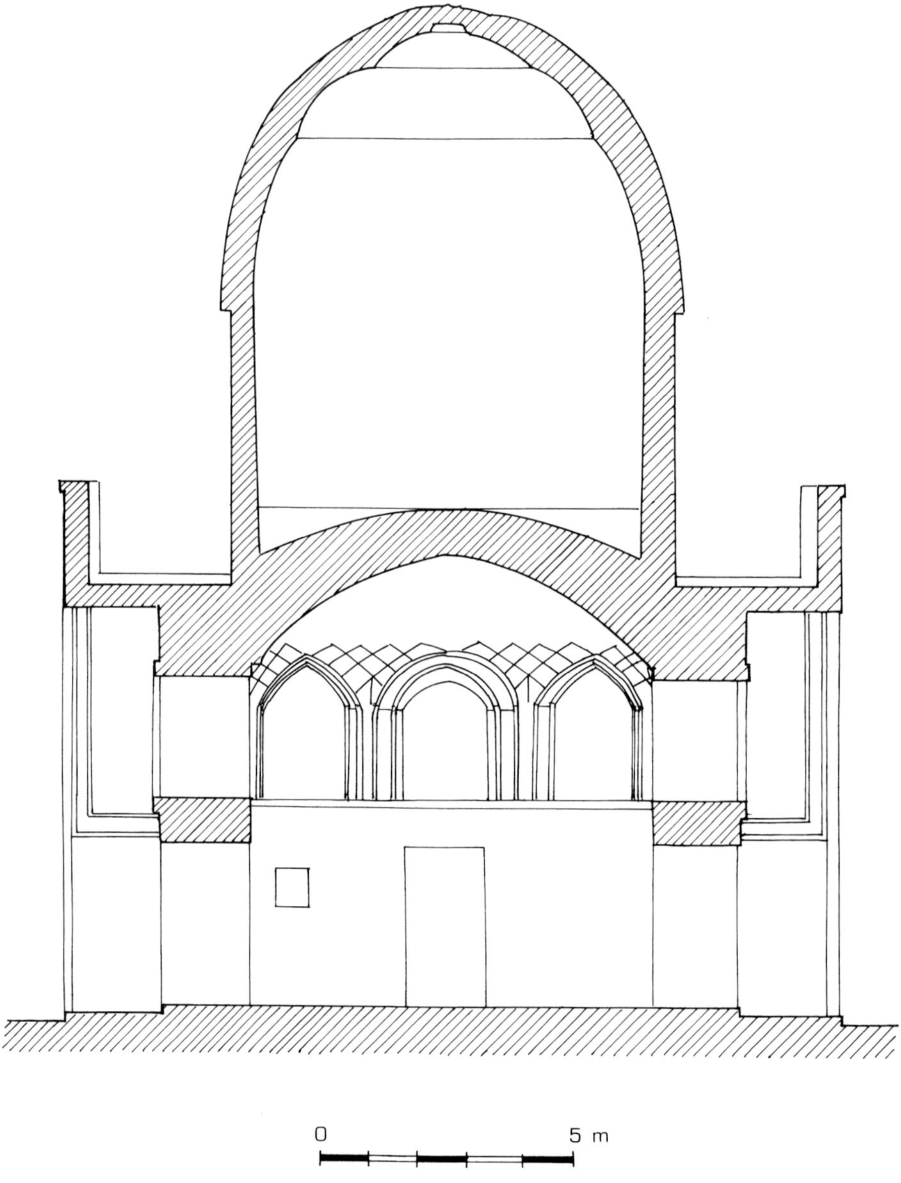

Fig. 87. Delhi, Sabz burj, cross section.

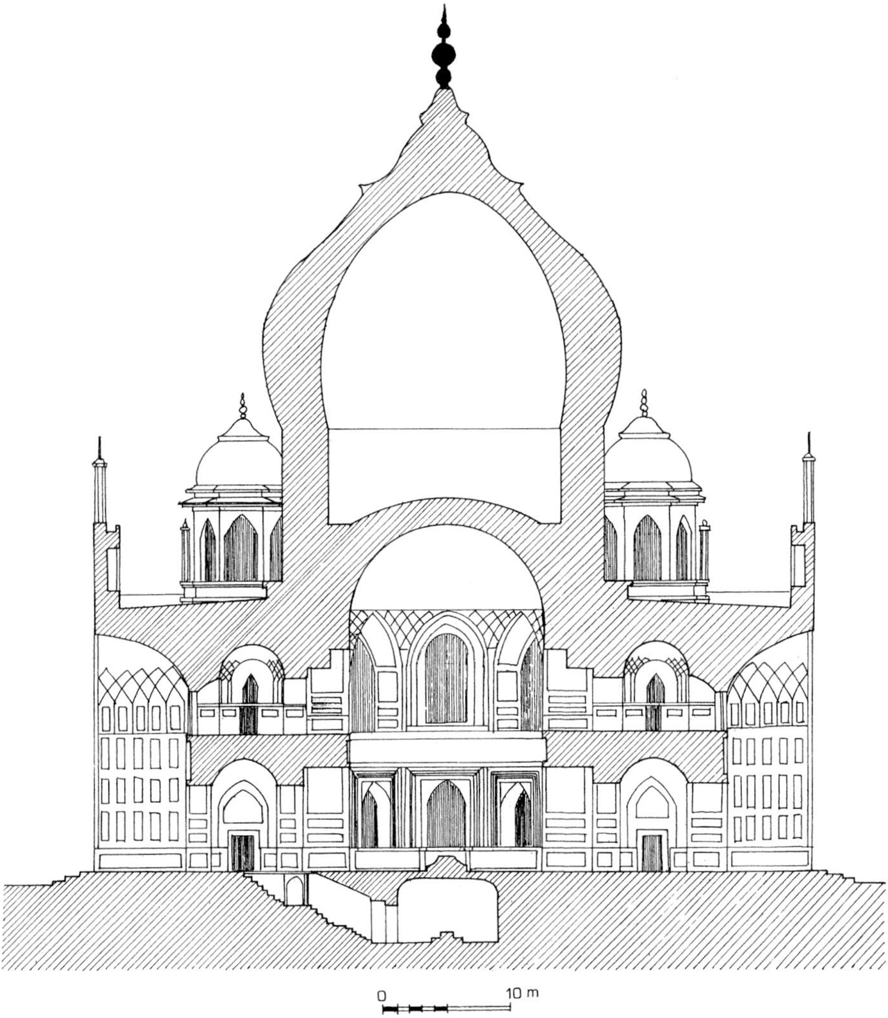

Fig. 88. Agra, Taj Mahal, cross section.

Fig. 89. Candi Gunung Gangsir, Java, detail of décor set into the brick bond of the base of the sanctuary wall; the different elements modelled into the terra cotta are hatched.

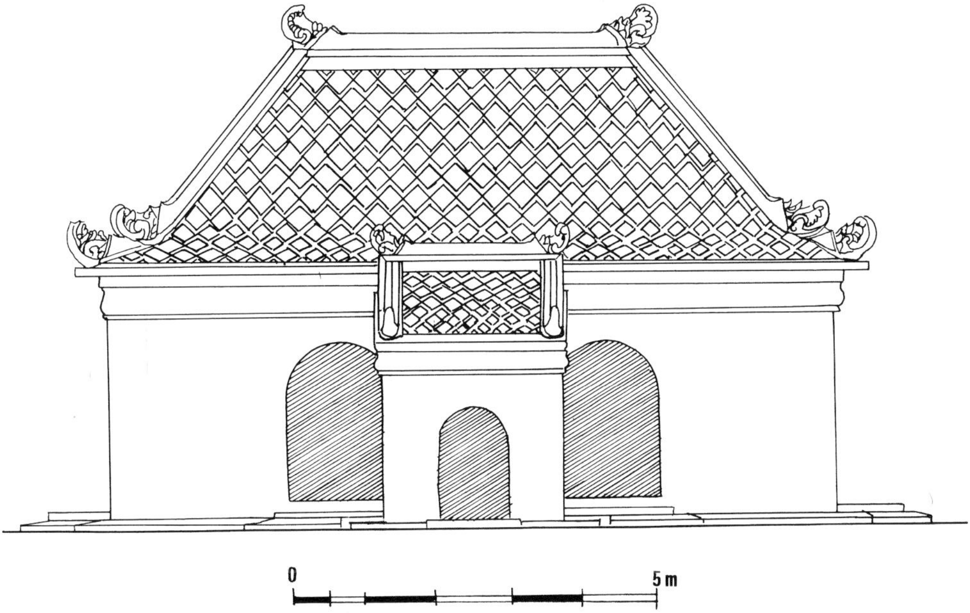

Fig. 90. Yogyakarta, Taman Sari, elevation of pavilion 14.

Fig. 91. Yogyakarta, Taman Sari, elevation of pavilion 8.

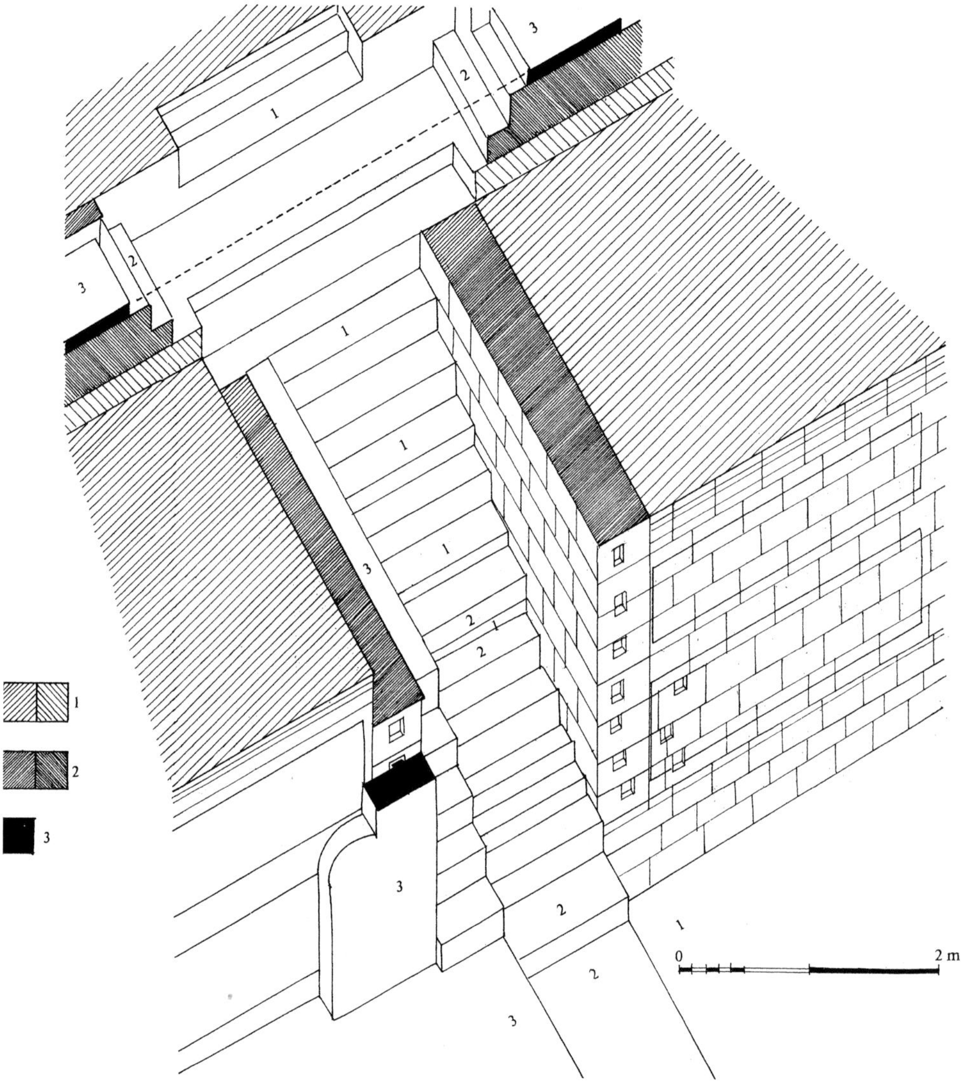

Fig. 92. Borobudur, repair of the second construction step, northern staircase from the first to the second gallery.

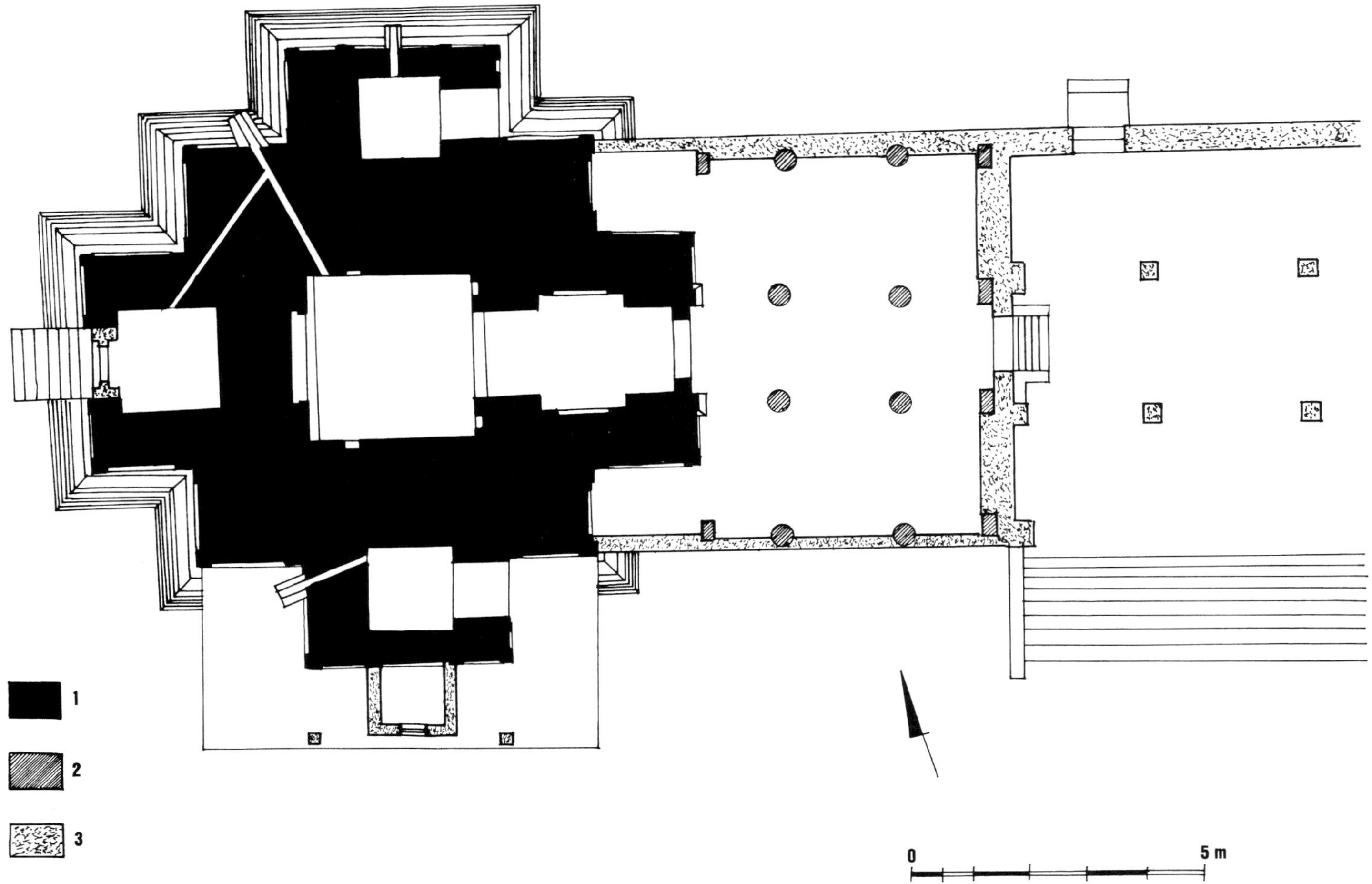

Fig. 93. Panamalai, constructions added to the primitive plan.

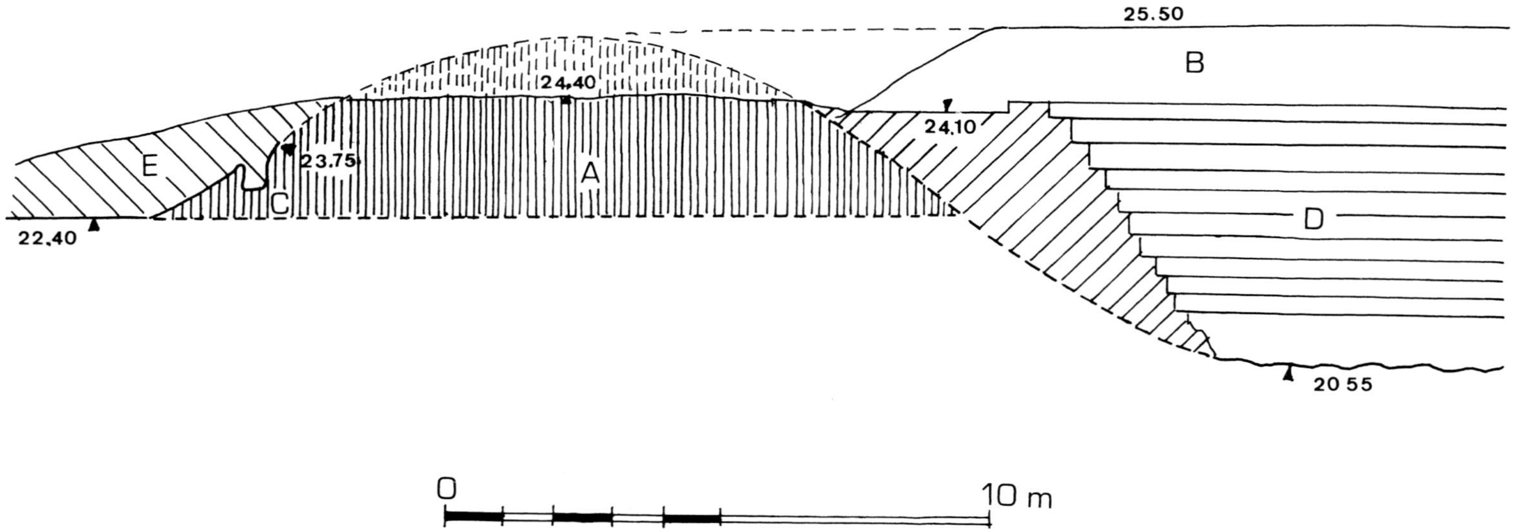

Fig. 94. Transformation of the Sras Srang, Cambodia. A, first state of the western excavation; B, first state of the northern excavation; C, canal of water distribution; D, second state of Sras Srang; E, excavated material after arrangement of second state.

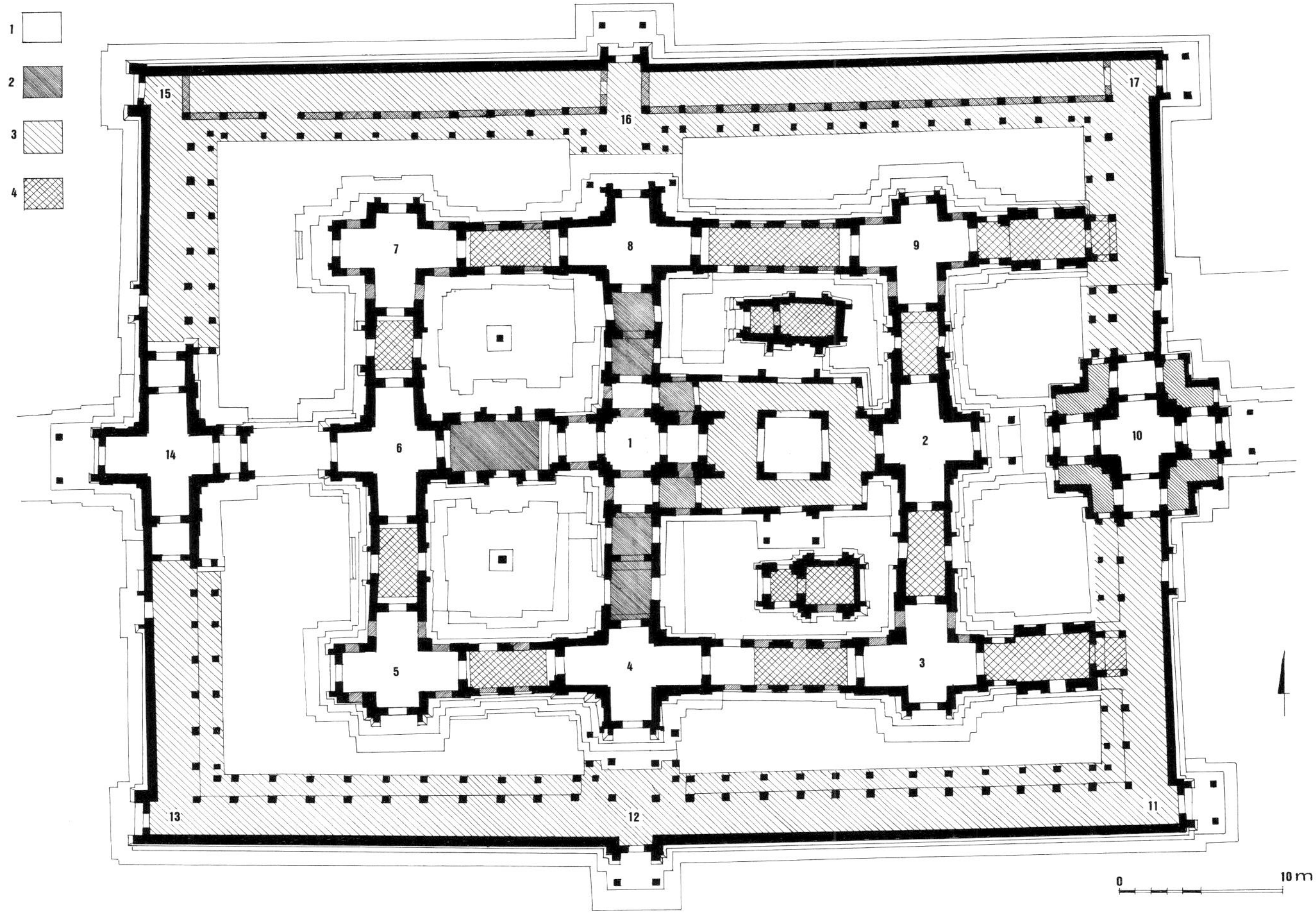

Fig. 95. Bantay Kdei, Cambodia, plan of different construction steps of the temple.

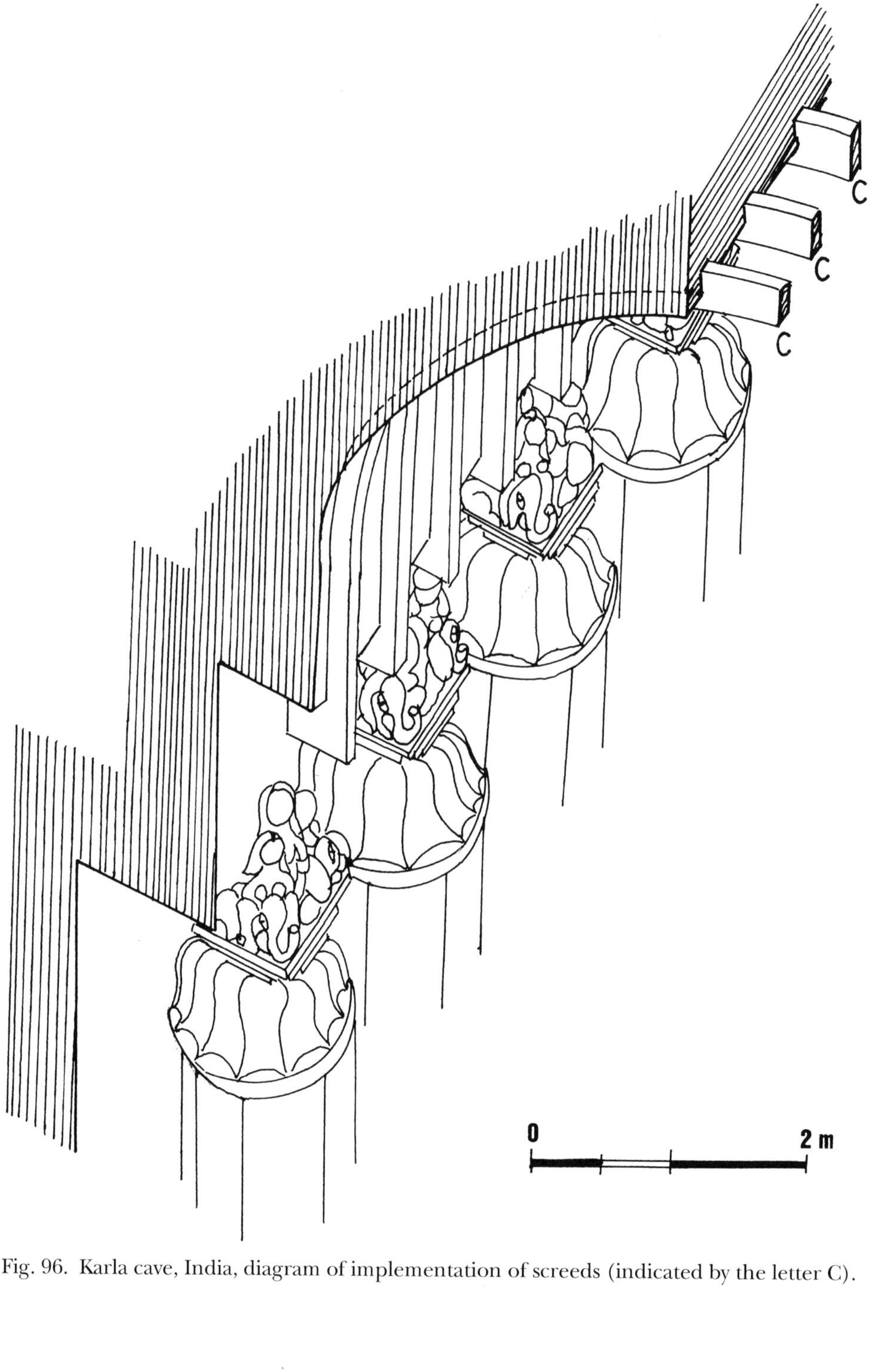

Fig. 96. Karla cave, India, diagram of implementation of screeds (indicated by the letter C).

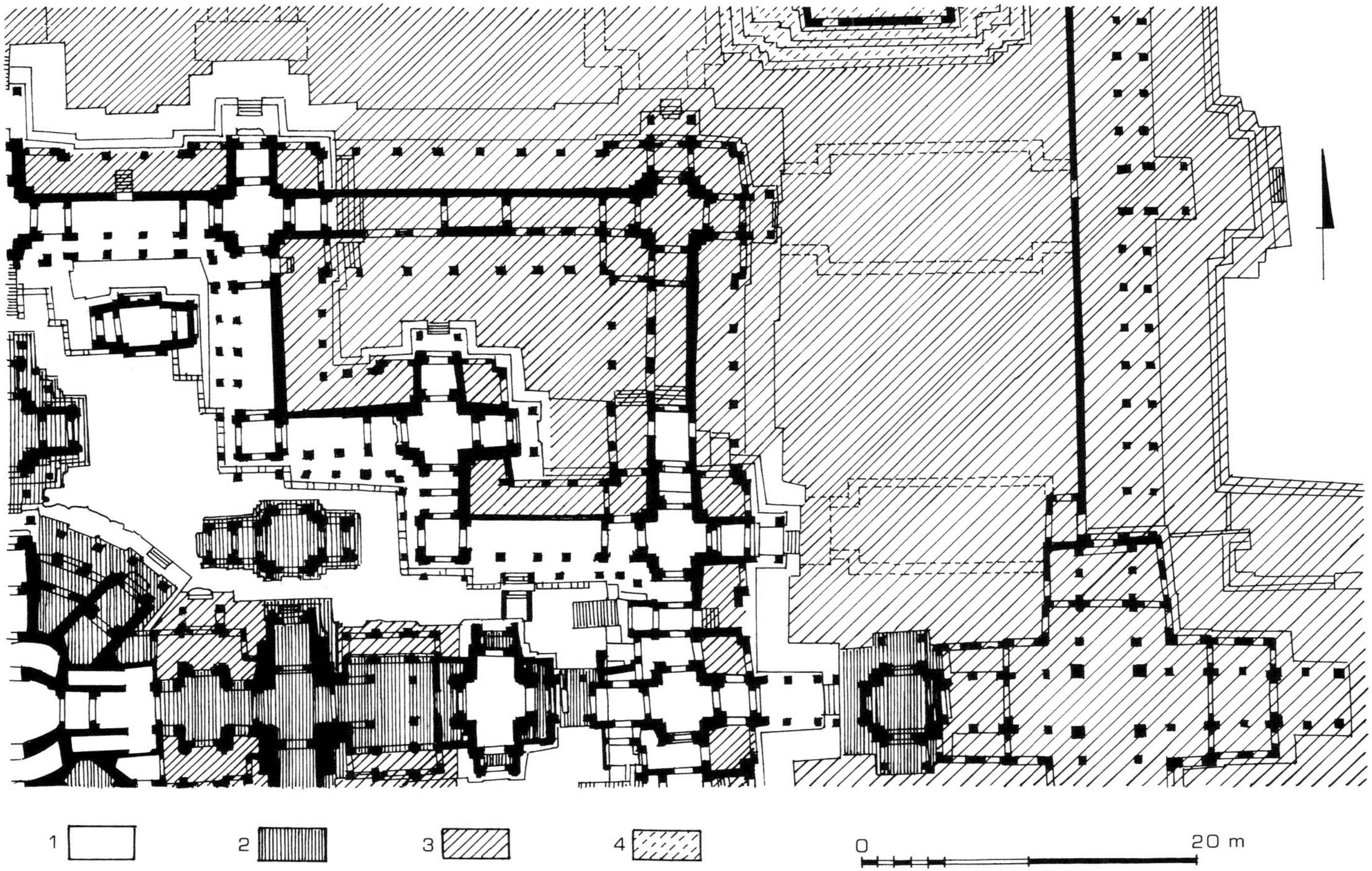

Fig. 97. Bayon, partial plan of north-eastern quarter, showing the four construction steps (marked 1, 2, 3 and 4). The structures indicated by dotted lines represent the galleries constructed after the third step and destroyed before the last step.

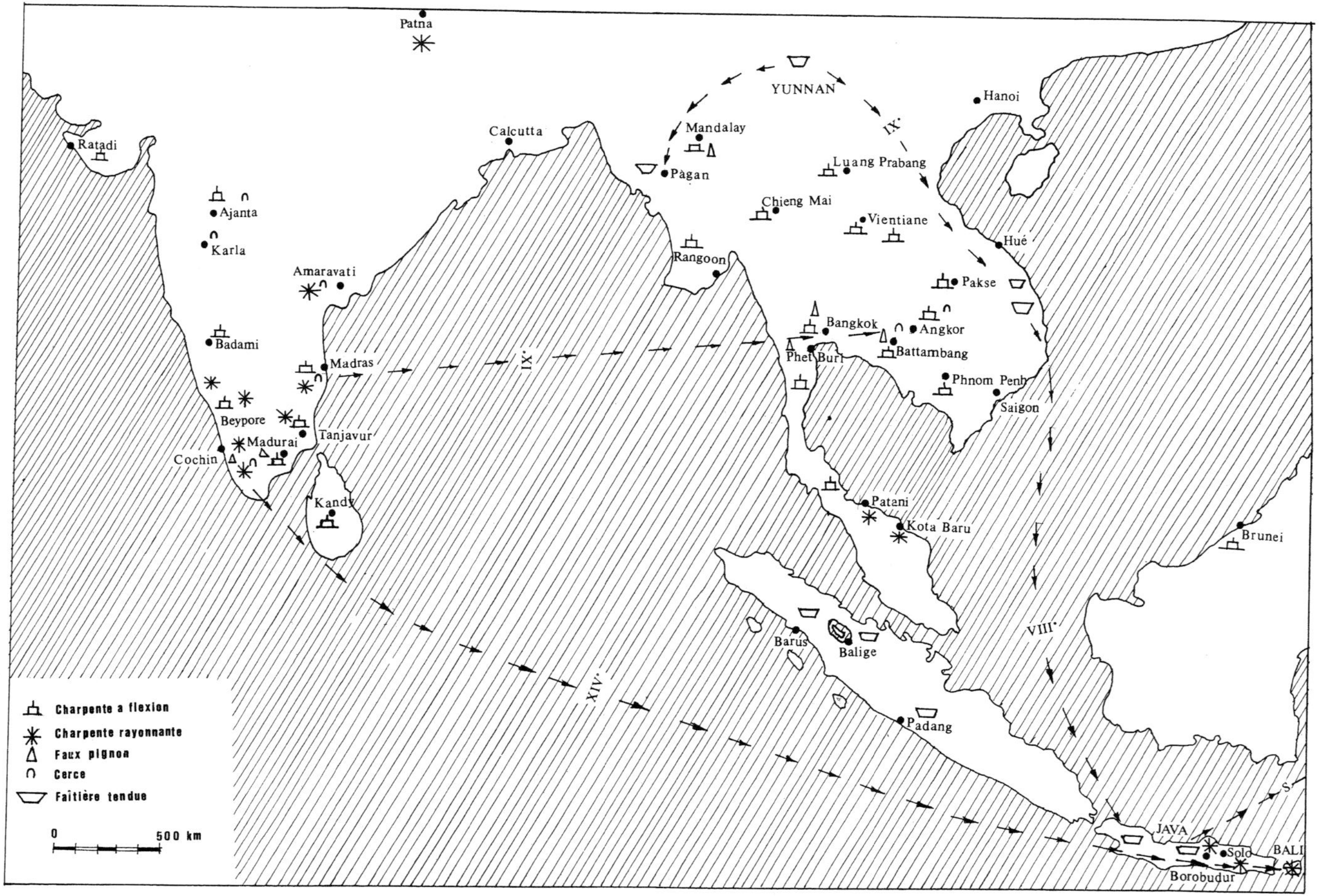

Fig. 98. Schematic map of southern Asia showing the breakdown of different carpentry techniques during the 13th century.

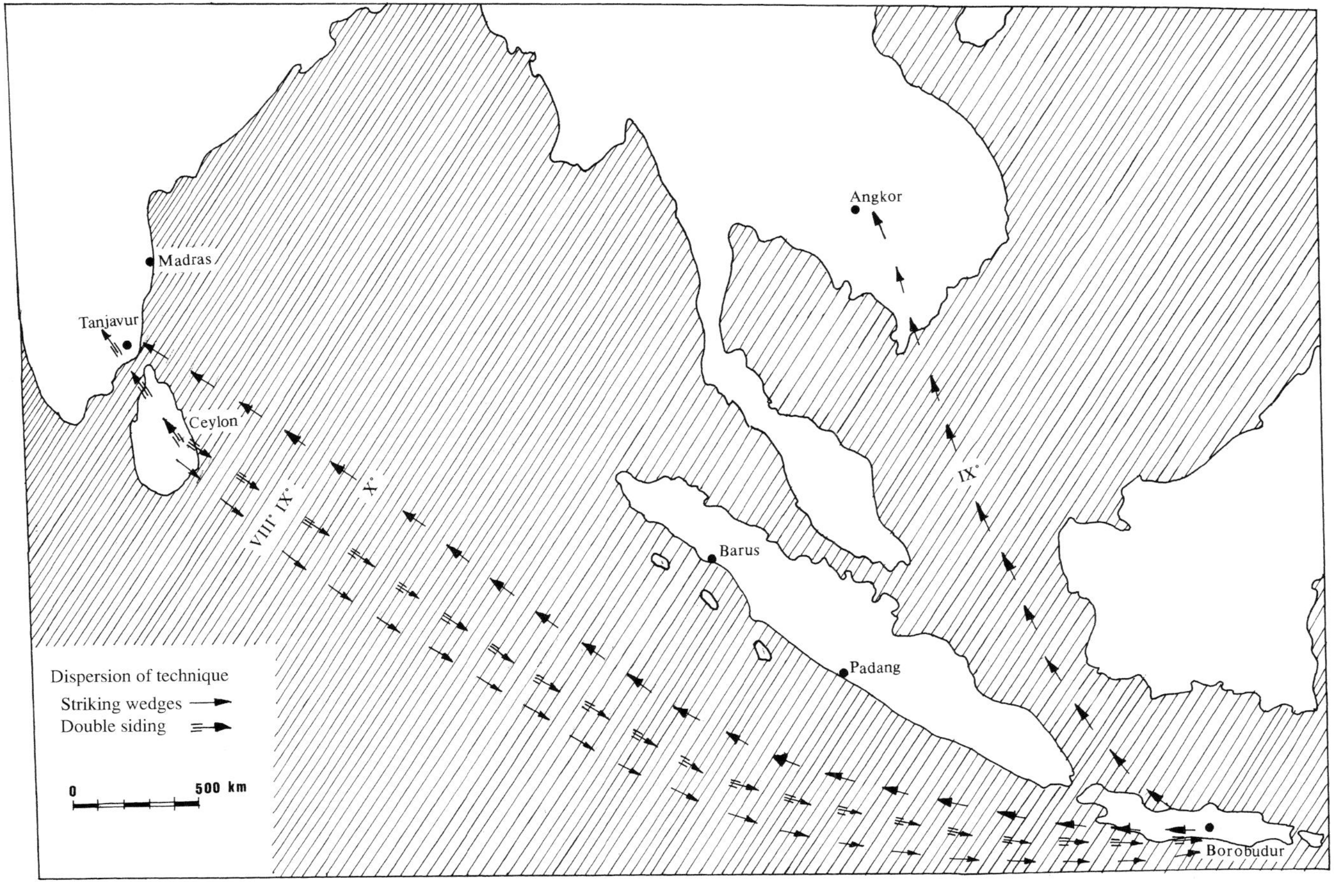

Fig. 99. Schematic map of the dispersion of the techniques of striking wedges and double siding throughout southern Asia.

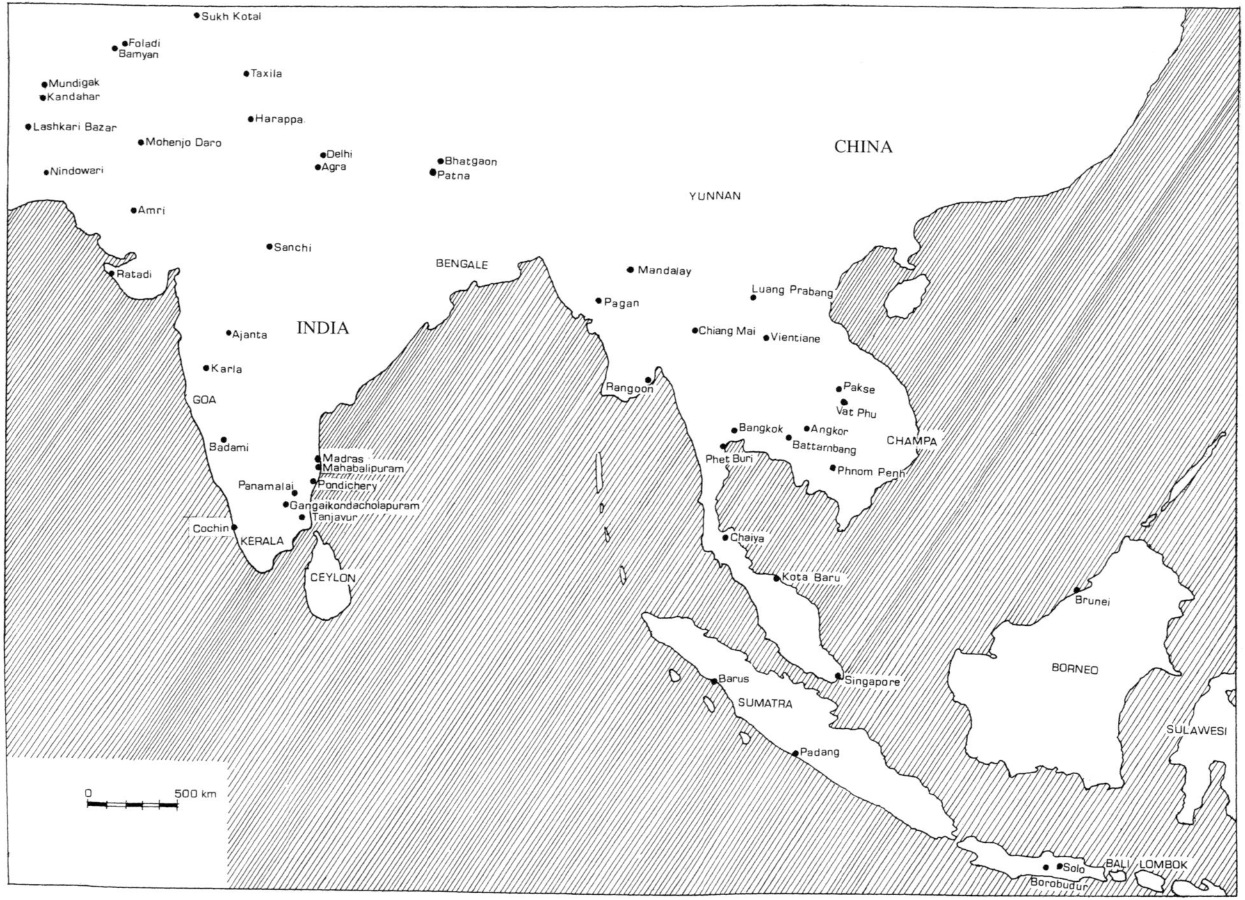

Fig. 100. Southern Asia, location of principal sites cited in this work.